Branes of Gravity

The Structure of Gravitational Force

Branes of Gravity

The Structure of Gravitational Force

H.C. Huang

Bellevue, Washington

huanggravity@gmail.com

Library of Congress Control Number: 2010916154
ISBN: Hardcover   978-1-4568-0503-6
                  Softcover   978-1-4568-0502-9
                  Ebook   978-1-4568-0504-3

To order additional copies of this book, contact:
Xlibris Corporation
1-888-795-4274
www.Xlibris.com
Orders@Xlibris.com
89287

The Flow

There is a long, long thin line:
that is my sorrow.

The Sorrow

From the lofty heights, I dropped
To the white peaks of thousands,
displaying their ever-resting silence
and solemn loneliness;
winding my way down into the gruff sea,
I tasted the bitter brine of the endless.

The Endless

Things circle themselves as the endless result . . .
Shatter the circle;
the will can free one's self.

—Hung C. Huang
May 2010

CONTENTS

FOREWORD

THIS BOOK, *BRANES OF GRAVITY*, follows my earlier volume, *A Simple Unified Theory: From Magnetism to Gravity*. This author disagrees with some of the current beliefs in theoretical physics. For instance, it is commonly believed that a perpetual machine is impossible. But this author thinks that our entire universe, galaxies, solar systems, and atoms are all, in effect, perpetual machines, whose motion still energizes everything we can sense.

A well-known conjecture states that, if the Sun were pulled out of our solar system, the Earth would continue its usual orbit until the Sun's gravitational wave, traveling at the speed of light, reached the Earth minutes later, thereby causing the Earth to veer off from its original orbit.

But that could never be the case. The Sun and its gravitational field—including our Earth, Moon, and all of the other orbiting planets—would simultaneously be pulled away together. Any reader familiar with the gyroscope understands that, the faster the inner wheel turns, the harder it is to alter the direction of the gyroscope. When the gyroscope, like any terrestrial body, moves, its gravitational field is not left behind.

With regard to other questionable assumptions, I leave it up to the reader to consider and discover.

—Hung C. Huang
Memorial Day (May 31), 2010

PART V

Negation

Section 1.
On the Principle of Equivalence

As discussed in my previous book, *A Simple Unified Theory: From Magnetism to Gravity* (Part IV), gravity is an inertia field comprised of multiple, small-scale inertia fields, which together congregate into the congruent field known as gravity on a grand scale. Thus, an inertia field or gravitational field is always curved. But space itself is neither curved nor non-curved because its existence is conceptual, meaning that it does not interfere with any shape, material, or energy, yet is simultaneously ever-existing anywhere.

In this case, how could space force a beam of light traveling in space to curve? Obviously, space cannot curve the beam of light because of its non-interference with anything. But since the gravitational field is always curved, the gravitational force will bend the light to curve in its field. Also, since the gravity fields are ever-changing, the path—and, in some cases, even the direction—of light traveling in the curvature is also ever-changing. We can imagine, for example, a beam of light cast into our vision on Earth from a million light years away; the light beam may not be a simple curved one. Instead, it could have been irregularly curved countless times before reaching our view.

How can it be proved? In our homes, for example, we can move any piece of furniture from one place to another; in so doing, we experience the gravitational force effect, that is, the effect of curved gravity, but not of curved space. Thus, the furniture is not controlled by curved space; rather, the space conforms to the furniture. The same phenomenon holds true in extraterrestrial space. Curved motion particles, such as photons, are affected by gravity's curvature. However, the gravity phenomenon is not controlled by curved space; otherwise, curved space moved from one place to another should remain the same. But this would not be the case unless the entire terrestrial system were moved. Thus, it is clear that the always-curved gravitational field of the planetary system, not curved space, forms the curved passage of the photons (see Part II, Section 3). The gravity field is ever-changing; but is the space curvature ever-changing? The answer would be "yes" if space had its own energy. But in relativity, space is curved, as if already fixed, and thus is not ever-changing.

A well-known experiment illustrates the principle of equivalence. Let us imagine a person with laboratory equipment riding in space in an elevator-like enclosed chamber, which is pulled upward by an attached cable (Fig. m1). The rate of acceleration is exactly that of the freefall speed on Earth. The person inside the chamber cannot distinguish whether he is standing on Earth or flying in space. If a beam of light passing through a hole in one wall of the chamber were cast on its opposite wall, then the person, using the laboratory equipment, would experience the beam of light as bent. Based on this example, the theory of relativity predicted that a star's light passing by another star's gravity field would bend. When we look at that star's light from the Earth, it appears that the star's original position has changed. According to this theory, acceleration is equivalent to gravity.

Now let us suppose that, instead of pulling the chamber upward, we swing it (Fig. m2). The person's experience of the bent light would be identical to that in the accelerating motion example (Fig. m1). However, he would experience a different feeling, being unable to stand still. We can also consider a third case, whereby a long rod is used to lift the chamber floor up (Fig. m3). The person would experience the bent light the same as in the previous two cases, but he himself would experience yet another feeling.

Let us now imagine all three chambers, with a person inside each, placed side-by-side, with floors level to one another. Let us imagine a hole drilled into the wall of each chamber where the light casts on the opposite wall so that the light passes out of each chamber. Next, let us imagine that the respective motions of the three chambers are activated simultaneously (Fig. n). All three persons inside their respective chambers would perceive the light bent in the same curvature (from the light's entry through the chamber wall on one side to its exit through the hole drilled into the opposite wall). If an outside observer were present, he could detect the true situation in a split second: He would see a light beam from its source passing through all three chambers and continuing in a straight line; that is, the light would not actually bend.

Among the three isolated cases (Figs. m1, m2, and m3), only in the first one (Fig. m1) does the person's feeling match the fact that light does bend while passing through Earth's gravity field. In the other two cases (Figs. m2 and m3), even though the light bends, the person does not feel that this is related to the gravity field.

Fig. m

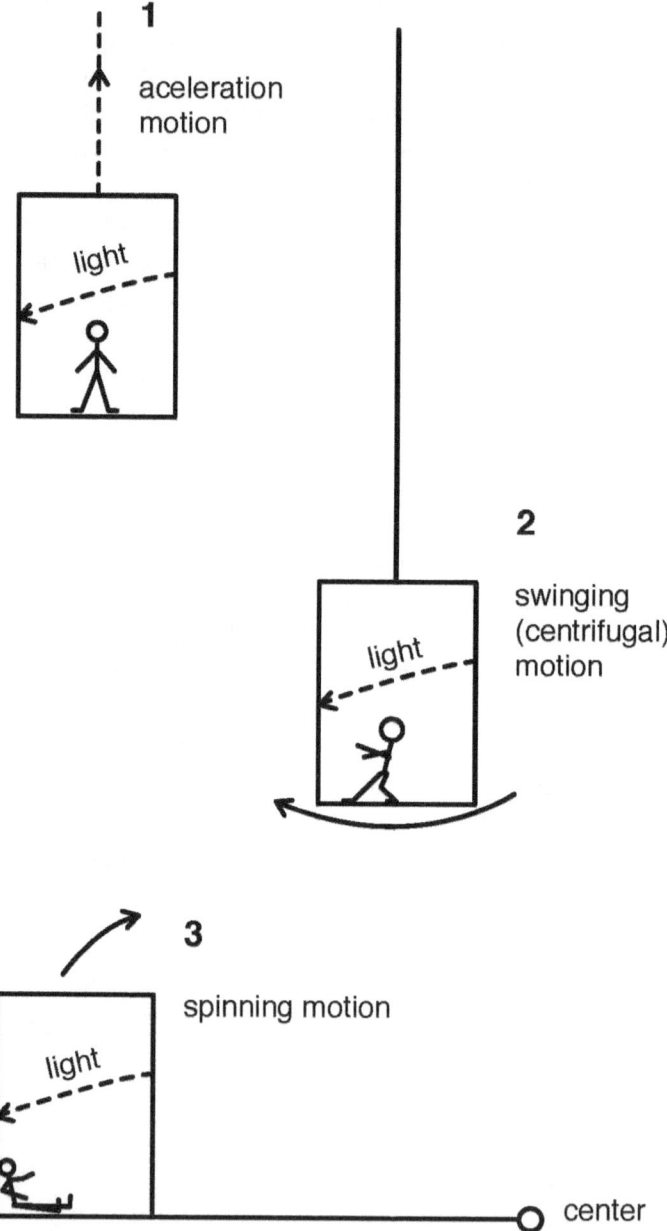

1

aceleration motion

light

2

swinging (centrifugal) motion

light

3

spinning motion

light

center

Fig. n

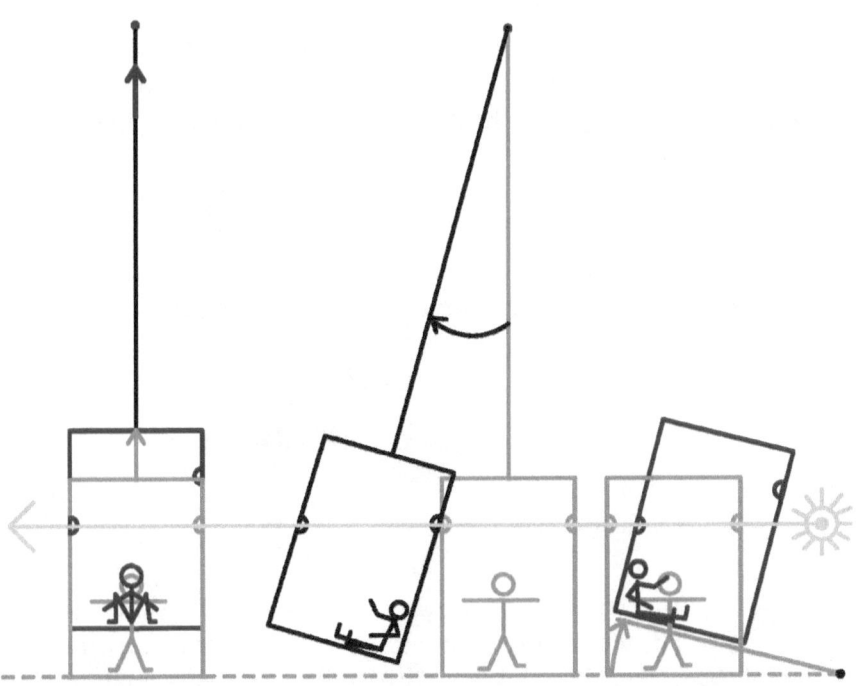

observer

Imagine that we place the Fig. m1 chamber in the Earth's gravity field in orbit, and the beam of light passing through a hole in the chamber wall casts light onto the opposite chamber wall. The person sees that the light bends, and feels he is floating. Next, if we initiate a force to push the chamber down at freefall speed toward the center of the Earth, the person sees that the light passing through the hole travels in a straight line, provided that the chamber does not spin. Again, the person feels that he is floating. However, the outside observer will see the light bend in Earth's gravity field. In yet another example, we place the chamber upside down in an orbit synchronized with the Earth's spin, making the person feel that the chamber is at rest. Of course, the person inside is floating, but the light passing through the hole bends toward the ceiling.

In short, a person's feelings cannot be applied to physics principles. In this case, the principle of equivalence cannot be inferred from merely one example involving feelings.

Section 2.

Is Uniform Motion Not Absolute?

Uniform motion is absolute.

In 1904, one year before Albert Einstein presented his Special Theory of Relativity, Henri Poincaré had already pointed out his "Relative Principle." He clearly expressed that a new dynamic theory was needed to replace the major role of Newton's theory. He also pointed out that we had not yet found any method to know whether we are in uniform motion or at rest.

In 1914, Albert Einstein published a paper rejecting the idea of absolute uniform motion, based on his Special Theory of Relativity. In it, he postulated that no observation made inside an enclosed chamber, even if turned into the most elaborate physical laboratory, could determine whether the chamber were at rest or moving along a straight line with constant velocity.[1]

Assuming that claim to be true, let us suppose there are two identical chambers at rest in space with a person inside each (Fig. o1). Seconds pass and nothing happens. Then let us suppose a case where the same two chambers are in a straight-line motion, moving toward each other at a high speed (Fig. o2). Inside the two chambers, the two persons do not know whether they are at rest or in uniform motion. But after a split second, the chambers collide (Fig. o3). These two cases show that we cannot apply the principle of equivalence of mass and energy to explain the different outcomes in the two cases.

With regard to the body with constant speed in a straight line, where does energy reside in this moving body? That part will be discussed in Part VII, Section 4. But first, we need to apply Einstein's case of using in a single chamber with a person inside it equipped with physical apparatuses. What kind of method could he use to determine whether he is at rest in space or in constant motion in a straight line? If that could be detected and it was discovered the chamber was in motion, what would be the speed and the direction? We need to know.

Moreover, the experiment based on the Special Theory of Relativity used only one chamber without extra conditions outside the chamber. Therefore, we too must use one chamber *in situ*, including inside it whatever laboratory equipment would be required to decide whether the chamber in space were at rest or in uniform motion.

---

[1]   See George Gamow, *Gravity* (Mineola, New York: Dover Publications, 2002), 117.

Fig. o

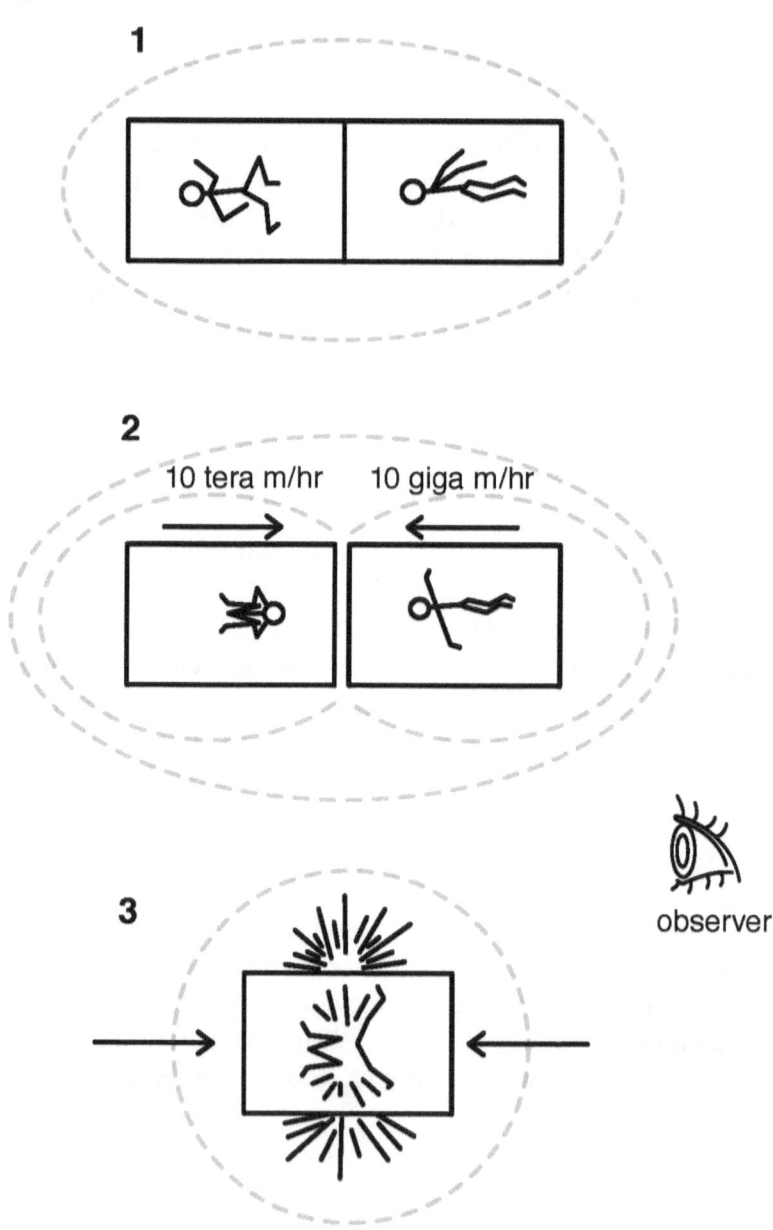

Let us imagine that the person inside this enclosed chamber in space is equipped with a device to shoot a beam of light from the exact middle point of one of the four walls (point LA) (Fig. p1). If the chamber is at rest, the beam of light will cast onto the exact middle point of the opposite wall (point A) (Fig. p1). This light beam will be parallel to the ceiling, floor, and two adjacent walls and perpendicular to the opposite wall, implying that the gravities of all four walls, plus the ceiling and the floor, are balanced. Obviously, the four sides of the chamber do not cause the light beam to bend.

However, if the beam of light casts onto the opposite wall at a point outside the exact middle point, the person will know that the chamber is in uniform motion. If the light has a red shift, he will know the chamber is moving in a direction away from the light source; conversely, if the light has a violet shift, he will understand that the chamber is moving toward the source of light.

If the chamber is in uniform motion, we also want to know its speed and direction. To determine these, we add another light source at the middle point of an adjacent wall (point LB) (Fig. p1). Thus, the two light beams cross exactly at the center of the chamber. Next, imagine a vertical line running through the center of the chamber ceiling down through the center of the floor (line z). Line z crosses the exact center of the chamber (point o) and is perpendicular to lines x and y. Thus, we now have coordinates x, y, z (Fig. p2).

Next, the beams of two sources of light in the middle of two adjacent walls are cast onto their opposite walls. If the light cast onto the opposite wall comes to rest away from the center point, this confirms that the chamber is moving. In this case, we draw a line to connect the center point of each wall (A or B) with its actual rest point (A' or B') (Fig. p2).

Fig. p

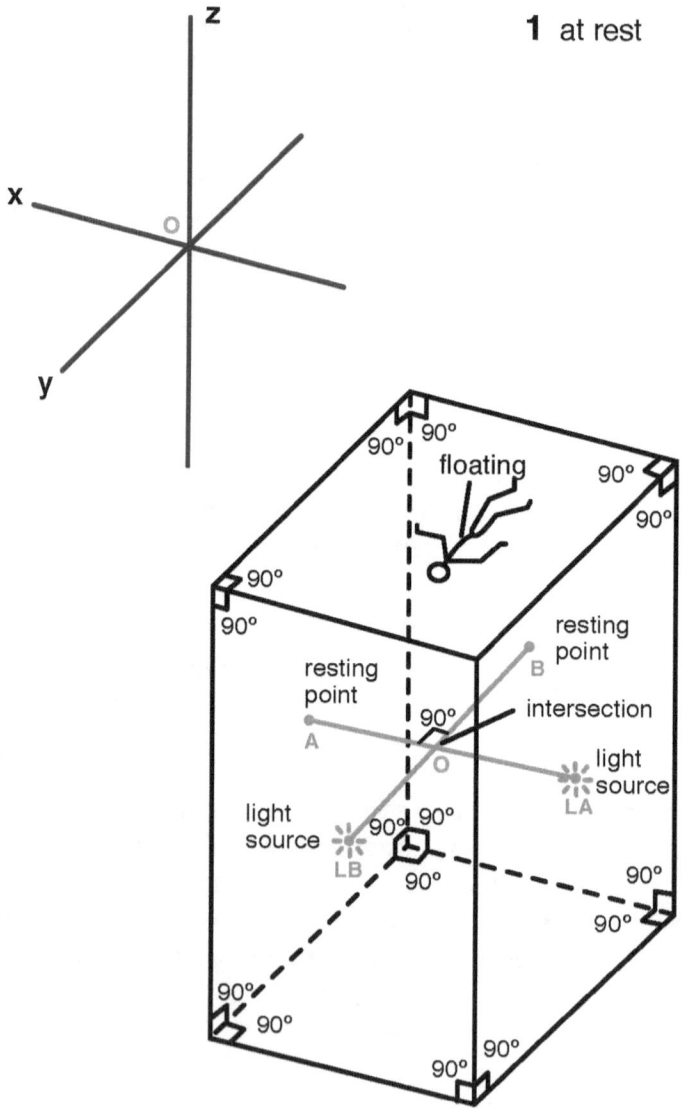

1 at rest

Fig. p

2 in uniform motion

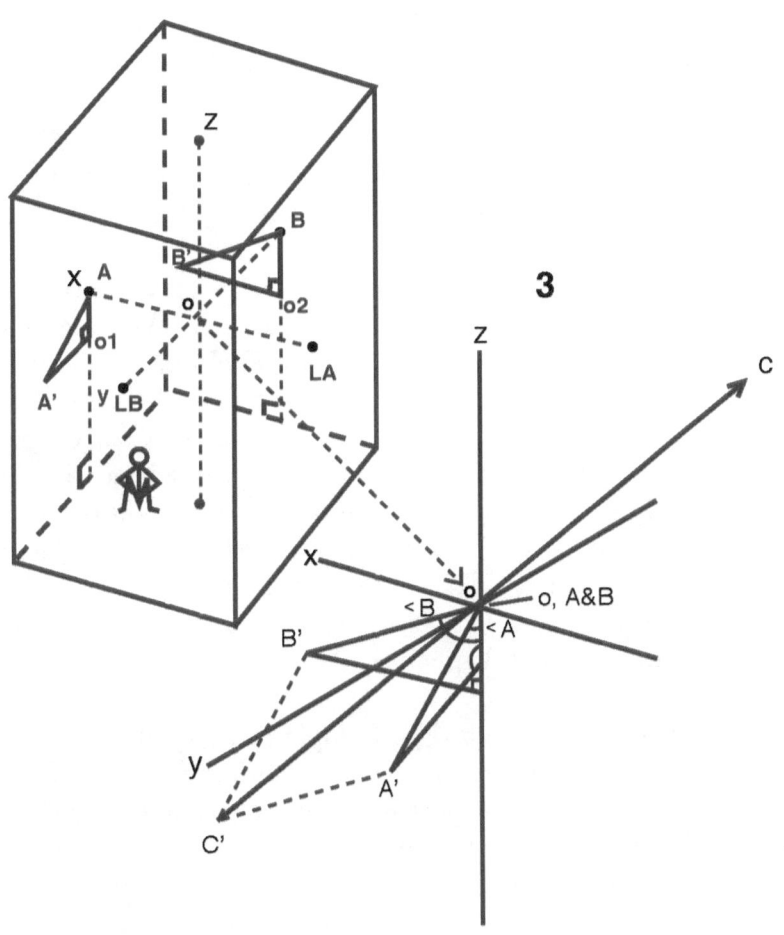

Since $\overrightarrow{c'o} = \overrightarrow{oc}$, $\overrightarrow{oc}$ represents the
speed and the direction of the chamber
traveling in space in uniform motion.

From point A or B, we then draw a vertical line down to the chamber floor. Then, from the actual rest point (A' or B'), we draw a line parallel to the floor, which crosses the vertical line at point $o1$ or $o2$, respectively. In both cases, we now have a right triangle ($A, A', o1$ or $B, B', o2$). Next, we move the plane of each triangle parallel to its wall and opposite wall so that the right angle side rests on line z (i.e., points A and B rest on point o, and points $o1$ and $o2$ rest on line z). Since the wall of point B is perpendicular to the wall of point A, the right triangle $B, B', o2$ must be perpendicular to the right triangle $A, A', o1$.

We now draw a line from point B' parallel to line oA'. From point A', we draw a line parallel to line oB' in the same plane. The point at which the two lines cross is C'. In this way, lines $B'C'$ and oA' are parallel and of equal length. Similarly, lines oB' and $A'C'$ are parallel and of equal length. Now we know that line oB' is vector oB' and line oA' is vector oA'. We then link point C' to point o diagonally in this parallelogram ($oB'C'A'$). We know that the diagonal line from C' to o is the vector of vector $B'o$, plus vector $A'o$, which is now vector $C'o$. As Fig. p3 shows, vector $C'o$ is located in the lower quadrant (III). We then extend vector $C'o$ in the same direction to point C so that vector oC equals vector $C'o$. We see that vector oC is located in the upper quadrant (I). Using this method, the person inside the chamber can find a vector similar to vector oC to determine the chamber's direction and speed.

As Fig. p2 shows, both right triangle LA, A, A' and right triangle LB, B, B' measure the distance from the respective light source points to their rest points. Since the speed used is the speed of light, both vector LA, A' and vector LB, B' represent the time required from the respective source points to their rest points. Thus, we have derived the speed value for vector oC.

This exercise thus shows that uniform motion is absolute. Likewise, an accelerating motion is absolute. Any uniform motion with spin will be similar to the example mentioned in Part II (Section 3) (i.e., the photon spirals in a straight direction). Therefore, all motions are absolute.

Section 3.

Vectors and Momenta

The major features of the neutral charge field are adhesiveness, plasticity, conduction, and oscillation.[2] Owing to its oscillating nature, the neutral charge field maintains its momentum and energy within a plastic domain. Because of its adhesiveness, it can surround a particle. When particles simply spin, this field allows the particles to spin freely. When particles spin eccentrically, this field will cause a rhythmic vibration. When the motion of particles is irregular, this field will correspond by oscillating irregularly. If the oscillation volume exceeds the tolerance level of the total field of minute particles, the outer layer of the field will split open, the extra oscillating energy will be dispersed, and the unbalanced part of the field with charges will be pushed out. This phenomenon, known as radiation, is the atom's way of balancing itself intrinsically.

When an object is accelerating or in uniform motion, it will generate a vector of general momentum. If this object contains sub-objects with motion in different directions, those small vectors will have free motion only if their total quantity is less than that of the object's large vector. But if the total momentum of the smaller vectors approaches that of the large vector of the main object, the small vectors will experience resistance. This is the asymptotic phenomenon, previously explained in Part III. Thus, a gravity field to which force has been applied gains momentum and becomes an inertia field. In turn, an inertia field's shape changes from ball to egg-like (assuming that the original object is ball-shaped). The "head" of the moving object's inertia field will be pressed against the object, causing the inertia field to flatten, thus forming its vector potential. Obviously, the "tail" of the moving object's inertia field will be dragged behind, meaning its inertia field will be looser. This is an example of a ball-shaped object that contains

---

[2]   For more details, see H. C. Huang, *A Simple Unified Theory: From Magnetism to Gravity* (Philadelphia: Xlibris Corporation, 2006).

momentum, suggesting the moving object's vector (Part IV, Section 4, Fig. h).

All radiation is matter. Radiation could either be in the form of particles with inertia fields or neutral charge fields with their momenta spiraling in a straight or curved line; the eccentricity of the fields adds to the destructive range of their path.

Another situation is self-spin momentum. In this case, a motion's energy and momentum behave like a vector, with the exception that the directions are always changing owing to the neutral charge field's adhesiveness and plasticity. An example is the polarization of light. This occurs when a ray of photons with its surrounding neutral charge field passes through a crystal material, which causes it to travel in an elongated, instead of round, spiral path because the neutral charge field can preserve the status of the photon's rotation energy, which is its momentum.

Therefore, the momentum of any object in a uniform motion in space is partially contained in every atom's neutral charge field and forms into a total neutral charge field to surround the object. This subject will be discussed in more detail in Part VII, Section 4.

For atoms that have been congregated into an element, the original neutral charge fields of their individual particles (i.e., protons, neutrons, and electrons) will form a total large neutral charge field surrounding the element. When several elements congregate to form a molecule, an even larger neutral field (the charge is weaker here) is formed, which is less condensed than the neutral charge fields within the atoms. Similarly, when the molecules congregate to form an object, the neutral field surrounding the object is even less condensed than the neutral fields surrounding its molecules. Since an object is matter, a planet consists of many matters. Thus, a planet has its surrounding neutral field, which is even less condensed than that of any of its objects. The planet's surrounding neutral field is its gravity field. Between each degradation interval, the volume increases enormously, causing the color force to degrade into gravitational force.

The belief that the gravitational force is too weak to compare with the color force does not hold true entirely because of the enormous difference between the volumes of these forces' domains. Needless to say, in the gravity field, each brane repels its neighboring branes; the counter-repelling force between the branes weakens the gravity force's strength.

For an object's motion to revert to its at-rest status, a force equal to its vector is required, meaning that the original vector of the object, which contains momentum and force, must be cancelled. If an object is at rest and a new vector is created or two vectors are combined, the object will form either a uniform motion or a spiral, straight-line motion. This explains why many particles in the universe, including the motion of photons, as well as their own angular momenta, also contain either a larger or smaller spiral motion. In other cases, some of the particles in motion carry their eccentric angular momenta fields, when affected by the surrounding forces, which cause their travel routes to curve.

The neutral field attached to each particle holds partial momentum of the particle's motion.

Section 4.

Theory of Gravitation

When any two terrestrial bodies approach each other, the cause is their gravitational pull. From Newton to Einstein to the present day, the attempt to better understand the nature of gravity has been a central concern. A recent theory of gravity introduces the concept of "fabric" as a curved space resulting from the curved movement of terrestrial bodies. The fabric will curve according to a body's mass. But the question is this: What is the direct attracting force to pull a terrestrial body down vertically? Another attracting force field must exist under the fabric to cause it.

Fig. q shows a star and two of its planets, one nearby (planet m) and one farther away (planet s). Since the star is of greater mass, it will sink to a greater degree, causing the fabric, and hence space, to curve. Planet m, of much smaller mass, creates a smaller curve underneath itself, and its movement tends downward, rolling toward the star. But the centrifugal force of the planet's motion prevents it from falling into the star; instead, it orbits the star.

What holds the star and its planets up? Is it light? Photon particles or light waves cannot hold terrestrial bodies up. In Fig. q, what causes the greater mass of the star to sink deeper and smaller masses of planets m and s to sink less? If gravitational force is the cause, then extra gravitational force must exist vertically under the entire system. One cannot logically explain the existence of gravity via the existence of another gravity, which exerts a vertical force underneath it.

Fig. q

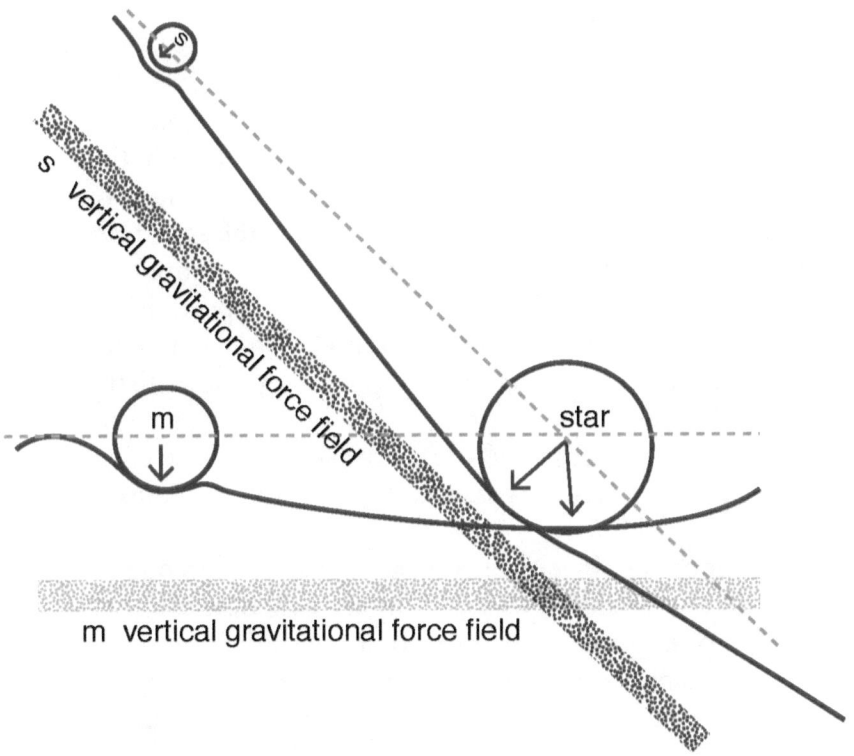

s vertical gravitational force field

m

star

m vertical gravitational force field

In Fig. q, planet s and planet m are in different planes of fabrics with different angles orbiting the star. Would not the two fabrics interfere with each other or space (represented by fabric) split into two curved existences simultaneously? How many vertical gravity fields would be needed to explain every planet in the solar system without interference between them?

This exercise shows that a fabric woven by light cannot exist because the theory is self-contradictory.

Section 5.

Contraction of Matter

In 1889, two years after the Michelson-Morley Experiment, George Fitzgerald still postulated that the length of a body traveling through the aether will change, and the changed quantity will relate to the speed of the body divided by the speed of light squared. Following Fitzgerald's claim, Hendrik Lorentz postulated that the traveling body will contract toward the speed direction, in accordance with the "contraction ratio" formula in Lorentz's transformation equations. However, Lorentz was dissatisfied with his own postulation, and Henri Poincaré rejected it as well. It was Albert Einstein who resolved the discrepancy in those equations, pointing out that mass, length, and time are not absolute.

That a moving object will contract toward its moving direction is a generally accepted concept. If the moving object were a spaceship, for example, a ruler or other measuring instrument inside would simultaneously contract. But the passengers riding in the spaceship would not perceive that anything had contracted.

Fig. r

1

at
rest

2

greatly
accelerating

3

acceleration ceased

(spaceship in
uniform motion)

But this book takes a different perspective. If the spaceship were to carry two identical rulers, one placed vertically and the other horizontally on the floor (Fig. r1), during high-speed acceleration, everything inside, including the rulers, would contract to about half their size toward the traveling direction; at that time, the upright ruler would retain its original length but would have become half as thick (Fig. r2). Once the acceleration ceased, a passenger could use the vertical ruler to confirm that everything is contracting (Fig. r3). Thus, the measurement and theory have inner discrepancies. Furthermore, if the spaceship made a U-turn to return to Earth, the spaceship and everything inside it would continue to contract, and the outcome would be irreversible.

One might imagine that an object in motion would contract toward its traveling direction. But this is not so.[3] For example, a gunshell pushed out from a barrel will not contract but will be compressed toward the traveling direction. Conversely, if an object were pulled, instead of pushed, toward the traveling direction, it would be dragged in the anti-moving direction because of its inertia. Thus, how a moving object is affected by acceleration depends on acceleration conditions.

Fig. s illustrates the motion of a spaceship and what happens to two types of bulk cargo and two passengers inside when the spaceship is at rest (Fig. s1), in acceleration (Fig. s2), and in uniform motion (Fig. s3). One type of bulk cargo is plasticable while the other is solid; each cargo is placed on a fixed, open-back chair with a frame. The two passengers stand facing each other, holding onto a fixed vertical pole (Fig. s1).

---

[3]   One may imagine that an object contracts toward the moving direction, taking that as the reason the gravity field causes the formation of a black hole; but this is not so. This is discussed in Part X (Section 3).

Fig. s

1

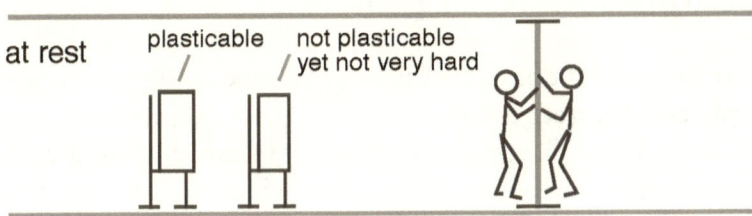

2

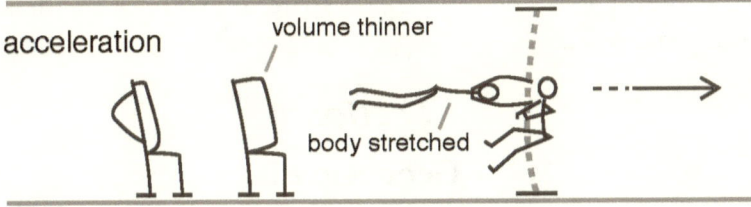

3

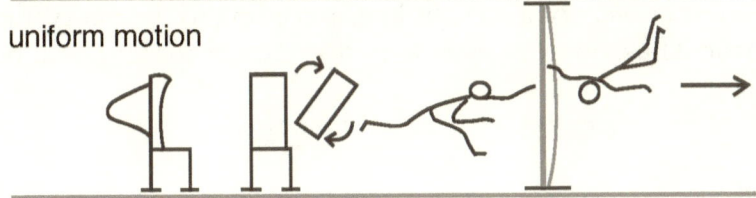

When the spaceship accelerates, the plasticable cargo is pulled backward through the chair's open back, while the solid cargo is compressed against the back of the chair. The passenger holding onto the pole facing the moving direction is pulled up and stretched longer, while the passenger with his back to the moving direction is compressed by the pole (Fig. s2). When the acceleration stops and the spaceship begins its uniform motion, the solid cargo recovers its original shape, the plasticable cargo (having been dragged during acceleration) remains that way, and the two passengers float (Fig. s3).

If any object contracts toward the moving direction or is simply compressed, it will somehow release a fractional amount of energy. Once the pressure of acceleration stops and uniform motion begins or resumes, the object will automatically rebalance its structure (but a fraction of energy or matter could already have been lost, in which case the status of the rebalanced object would not be exactly like that of the original one). A similar condition applies to an object pulled toward the moving direction (because of the energy loss, the rebalanced structure will not be the same as the original one).

Section 6.

Geodesic Line

According to the theory of the geodesic line in the gravity field, the space surrounding the Sun is curved. The planets' motion must follow that curved space because their orbiting route is the shortest curve, not a straight line, according to their curvature tensor. Thus, uniform motion in that curved space should not be true.

If true, then what is the source of the force that causes a planet to orbit? A planet cannot obtain any energy from space, no matter whether the space is smooth or curved. A planet that slopes toward the center of the Sun may freely choose its direction, either left or right, in which case the Sun's spin decides the direction of the planet's orbit. The curvature tensor is always changing because many planets possess their own curvature tensors. Furthermore, in this solar system, the total orbiting energy of the planets is larger than the Sun's own spin. That is, it is the total orbiting energy of the planets that moves the Sun from its at-rest status into spin.

Under these circumstances, one cannot say that the Sun creates its surrounding curved space; according to the geodesic theory, that is the origin of its gravity. As illustrated in Fig. q (Part V, Section 4), if that were true, it would require a parallel and flat vertical force of another universal gravity field. Therefore, that the geodesic line is the shortest route to move possesses no meaning.

As explained in Part IV, it seems that space is curved, but this is not exactly true. The origin of the gravity field derives from the matter of many atoms, which contain the positive, neutral, and negative charges. These three charges generate the energy for matter to have motion, such as self-spin and orbiting. Since a gravity field is always changing, so is its curvature. The individual gravity field of each terrestrial body affects the motion of its neighboring body without actually touching it; that is, a gravity field is the thinnest of matter. When a beam of light from a remote star passes through a gravity field, it may not necessarily pass through a fixed curvature, such as a geodesic line. But amid the twists and turns between and through those many gravity fields, a beam of light may encounter an empty space where it resumes its straight-line traveling route before reaching the observer's eyes on Earth. Because a light beam is composed of photons and a photon has mass, the light beam is flow of matter. As it passes through all shapes and sizes of gravity fields, the traveling route of the light beam will be affected by those thinnest fields of matter.

If one is really convinced that curved space can replace the functioning strength of gravity, then why still search for gravity waves or gravitons? That point of view contradicts the concept of curved space. If a gravity wave stands, then any particle traveling through it will have a speed in rhythm. If gravity is caused by exchanging gravitons, then one could not predict the number of gravitons a photon would have to carry during its traveling. If a photon did not need to carry any graviton, then the environmental gravity fields would have to supply gravitons onto it. Then imagine the situation: a photon simultaneously bombarded by gravitons from all directions and still having to throw them back to their original sources. Could this photon balance its energy of motion and maintain its speed of light? Unlikely.

The concept of forces based on particle exchanging will not always hold true in physics. A supplemental discussion of gravity is presented in Part VII.

Section 7.

Summary and Conclusion

Although it appears that the principle of equivalence holds true, the two cases are distinct, meaning the principle does not fit every aspect; therefore, the term *principle* does not apply.

All motions, including uniform motion, are absolute; moreover, it is possible to measure uniform motion.

Curved space cannot be used to explain all facts related to gravity. Furthermore, what force causes space to curve since space has no substance to receive such a force? Is the universe expanding because of the force of curved space? Curved space does not possess any force to change its own curvature.

An object in motion does not contract to its traveling direction. On the contrary, because of its own inertia, the object is pulled back in the opposite direction; the degree to which it is pulled back depends on the object's chemical structure. When acceleration stops and uniform motion begins or resumes, the object tends to rebalance its structure.

The direction of a planet orbiting a star is ever-changing, as is its inertia; the planet's speed is also ever-changing because the orbiting route is elliptical, rather than circular. Does this then mean that the geodesic line, the shortest distance, is also ever-changing? Does it also suggest that the curvature of space is not fixed, which contradicts the theory that the central star created the curved space, in which case the curvature surrounding the star should be round? But if a planet has its own curved space, which changes the star's curved space, what causes the frequent change in the planet's speed? Where does the energy required to change the speed and direction originate? Curved space cannot answer this question. Using curved space to explain gravity offers no force; rather, it seems the solar system always has forces and counterforces acting on each other. In short, a three-dimensional space produces no forces and no fourth dimension of time.

Fig. t

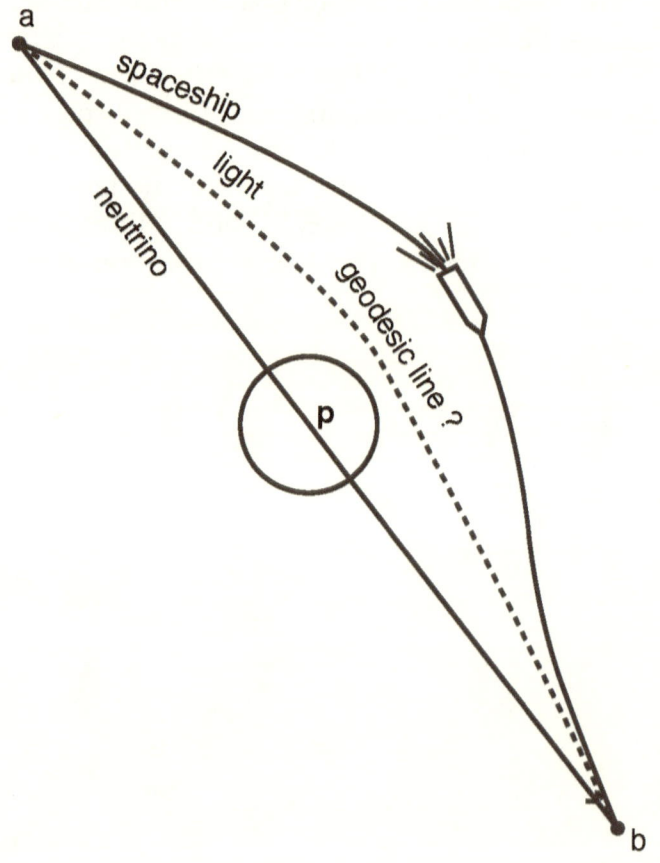

It is said that, when an object passes by a planet's gravity field, the geodesic line is the shortest distance. Fig. t presents three cases from point *a* to point *b*: (i) a light beam passing by planet *P*, (ii) a spaceship passing by planet *P*, and (iii) a line of neutrinos passing through planet *P*. The third case illustrates the shortest route traveled, while the second case shows the longest traveling route. Only the first case, the light beam, fits the meaning of the geodesic line.

PART VI

Special Review

THIS PART OFFERS A MORE IN-DEPTH DISCUSSION of the four main topics covered in the preceding book, *A Simple Unified Theory: From Magnetism to Gravity*, and further develops its key concepts.

Section 1.

Magnets

In Part I, Section 2, Fig. a showed that a magnetic line of force is a string with a positive nuclear quantumized charge, which is an extension of a lineal link of protons and neutrons, rather than the protons themselves (Fig. u1 is derived from Fig. a). A hydrogen atom functions like a miniature magnet, with its single electron revolving around its proton instead of oscillating.

Why does an electron sometimes oscillate and at other times revolve about the nucleus? The answer is that a proton spins, and the angular momentum of a proton's self-spin affects the neutral field surrounding it. For example, when heat is impressed into a hydrogen atom, the proton's spin causes its surrounding neutral field to oscillate; this, in turn, indirectly affects the electron's neutral field, slowing the electron's revolving speed into electronic oscillation. This status causes the hydrogen atom's magnetic force to weaken because the excessive neutral field from the heat has been impressed into the atom. If, on the contrary, the excessive neutral field is released or repelled—meaning that the atom is cooling down—then the atomic charge reverts to its original status, and the electron's orbit resumes its normal speed. This indicates that, when a higher-energy electron releases its energy, it leaps back to its original orbit.

A similar phenomenon occurs when a central axle spins in one direction: Its connecting outer body will move in the same direction (e.g., a helicopter). Of course, one could say that the reason is that the friction between the axle and the outer body causes the outer body to move in the same direction as the central axle. Yes, the neutral field does cause the friction created by the electrical charge to increase.

When an electron escapes from an atom, it carries along its kinetic energy in its own neutral field and maintains its spinning in the air. Because the electron's surrounding neutral field spins with it, the electron's domain is enlarged, thereby enabling a positive-charged antenna to more easily catch the electron as it passes by.

In Part I, Section 10, Fig. Super showed that quantumized charges, and hence a superconductor's magnetic strength, requires more explanation. How does quantumization of magnetic force happen?

When the temperature of a superconductor drops, all of the neutral fields in its molecules are reduced three-dimensionally. (For the same reason, liquefied air under very low temperature loses its viscosity to become superfluidity.)[4] The external force of the orbiting electrons, with their sweeping force, can induce the rest of the electrons to orbit freely in the same direction.

---

[4]   In 1924, Albert Einstein predicted superfluidity, based on Bose-Einstein statistics. In 1928, W. H. Keeson (1876-1956) discovered the superfluidity of liquid helium; similar discoveries followed in 1962 by Lev Landau (also on superfluid helium) and in 1996 by David M. Lee, Douglas D. Osheroff, and Robert C. Richardson (superfluidity in helium-3).

Fig. u

1

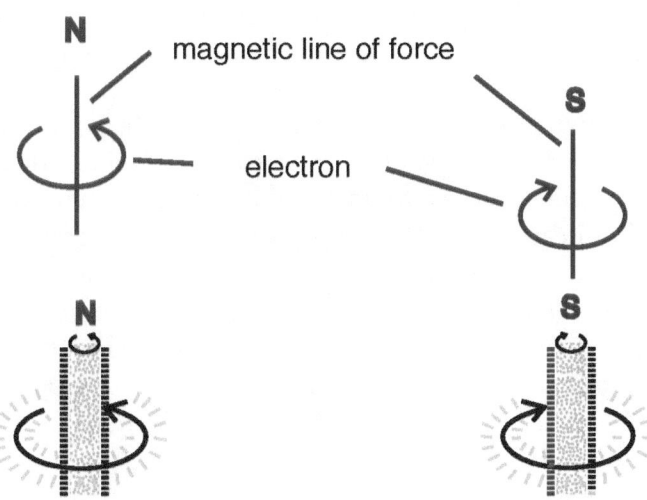

N

magnetic line of force

S

electron

N

S

2   quantumized
          magnetism

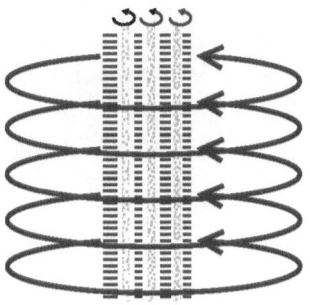

At the same time, the positive axles in the molecules gather together and push some of the electrons in their atoms into the outermost orbiting fields. Since the orbiting fields are perpendicular to the central positive axles, the freer orbiting motion causes the revolving electrons to flatten, enabling the neighboring electrons to move closer to each other and orbit in the same direction.

Simultaneously, the counterforce from the negatively-charged electrons forces the positive axles to extend their opposite ends into the air to form enlarged magnetic circuits, and carry with them the flattened, fast-orbiting electrons. In this way, the magnetic circuit possesses super-strong magnetic force. Flattened electrons can move at their fastest speed and minimize the gap between their neighboring electrons. One magnetic circuit can carry many more electrons. This is how the quantumization of magnetic force occurs (Fig. u2) (see also Section 4).

Section 2.

Photons

Part I mentioned that the "+" charge tends to be the center axle with self-spin and to connect with other positive charge axles. Like a magnetic line, it will connect vertically or loop with other magnetic lines. When a positron collides with an electron, both will break up into photons. Since the positive charge in the photons lacks enough force to maintain a straight-line self-spin—such as is found in the positive charge of a proton or the negative charge of an electron—the broken negative charge surrounding the positive charge in the center line will cause it to curve repeatedly. That is why photons are in the shape of a spiral and travel in a straight line (Fig. v).

Fig. v

When a photon vertically hits a target, especially one made of soft metal, a portion of the photon's energy may be absorbed by the matter, while the remainder may recoil from the target material. If a photon hits a material's surface at an angle of incidence and rebounds from it, then its angle of incidence will equal its angle of reflection, owing to its spiral spin, which enlarges its contact surface. When the photon hits the surface of the target material, frictional force will cause a minute quantity of the photon's neutral field to be stripped away, which, in turn, diminishes its frequency.

A photon is a particle of matter, and even X-rays and gamma rays are corpuscular streams of matter. There is a commonly held belief that, because a photon does not possess mass and the Sun causes its surrounding space to curve, a light beam from a remote star passing by our Sun will bend. However, this is not entirely so.

A light beam consists of photons, which possess mass, while the gravity field surrounding a star, such as our Sun, possesses the thinnest mass. Thus, when a light beam from a remote star passes by the gravity field surrounding our Sun, the Sun's self-spin will cause its gravity field to turn with it, which will affect the moving direction of the light beam. If the light beam passes by the side moving counter to its direction, it will curve and violet-shift its frequency; conversely, if it passes by the side moving in the same direction, it will bend less and red-shift its frequency.

Therefore, when a light beam passes by a star's gravity field, that star's self-spin speed will affect the light frequency.

Section 3.

Nuclear Surface Tension

The discovery of the simple fact that attraction and repulsion exist between electrical polarities engendered a great change for science and industry.

On Christmas Eve, 1938, Otto Hahn (1879-1968), Lise Meitner (1878-1968), and her nephew Otto Frisch (1904-1979) released a paper explaining that a nucleus, like a drop of liquid, has its nuclear surface tension to hold the nucleus together, as well as the repellent force from the charges of the protons; thus, the nuclear surface tension and the repellent force can be balanced. As the atomic number increases, they found that this repellent force strengthens; they calculated that, once the atomic number reached 100, the nuclear surface tension would be cancelled. Obviously, it is highly unlikely that elements with atomic numbers as high as 110 or even 115 could be found in nature; however, it may be possible to produce them artificially. Here we discuss why nuclear surface tension is related to nuclear-binding energy.

In the case of hydrogen, for example, the atom contains one proton and one electron. We know that deuterium has one neutron, while tritium has two neutrons. Why then does the proton allow either one or two neutrons to co-exist in the nucleus?

The strength of the proton's "+" charge is greater than that of the electron's "−" charge. To balance the strength of these electrical charges, the proton attracts a neutron's "−" micro-charges. At this stage, the proton still has some positive force, which is barely enough to attract a second neutron. But the stability of this status (one proton plus two neutrons) is fragile, causing some vibration in the nucleus. This book assumes that the force of the proton's positive charge in the nucleus is greater than the force of the electron's negative charge. Therefore, as mentioned in Part III, the neutral charge field must compensate for a proton's stronger positive charge by attracting extra neutral charge fields. Thus, the proton's corresponding electron will

35

sometimes be isolated by the neutral field, with a radius at a farther distance from the nucleus. The neutral field is easily affected by external forces, which causes oscillation. Hydrogen atoms contain a large volume of neutral fields and thus easily affect each other's oscillations, which causes fusion.

If a hydrogen atom were cooled to an extremely low temperature, its neutral charge field would contract outside its proton, causing the electron to change from oscillating into orbiting energy. The ratio for this proton magnet would be greatest since the electron's orbiting speed would be super fast.

If a hydrogen atom absorbs a neutron to become deuterium, its orbiting electron speed will decrease to become partially oscillating and orbiting. Since the deuton can absorb another neutron to become tritium, the triton will release some of the external neutral field. The orbiting radius of its electron, though reduced, can cause oscillation because the radius from the electron to the proton keeps changing. And this fast vibrating status occasionally changes from one of equilibrium into isochronous vibration, which indirectly affects the neutral field's strong oscillation. When oscillation peaks, the hydrogen atom's neutron, along with its neutral field, is released as radiation.

Even with one proton, the electron's orbiting does not glide smoothly. Therefore, in an element whose atomic number approaches 100, the stereochemical constitutions tend to lose some affinity at various points between the atom's protons and neutrons. The reason is that each proton can hold only one neutron firmly; a proton holding two neutrons can easily lose one of them. Helium (with two protons and two neutrons) again serves as an example. Each of the element's four sides has one proton and one neutron, which is the strongest binding affinity. Even though helium loses two of its electrons, its nucleus maintains its status of two protons and two neutrons, as an alpha particle, which is surrounded by a rather large neutral field.

This indirectly proves that nuclear-binding power, which we call either the strong or color force, is the binding charge of a proton that attracts a neutron. A proton bound with a neutron will reject the addition of a second proton; but adding another neutron to the deuton will result in a triton that attracts a second proton and binds together strongly. Clearly, the true color force derives from the positive and negative charges; thus, there is no need for gluons.[5]

In an atom, each increase of a proton with a neutron accumulates an extra positive charge and creates more phase angle. As the number of protons increases, their resonant matching phases will shift into greater instability. The more phase angles that are created by adding successive protons with their neutrons, the more cavities that will occur between them. Filling those cavities requires more neutrons, but that will cause the phase angles to shift from symmetrical to asymmetrical matching.

The outer nuclear surface tension of an element tends toward roundedness and smoothness and compresses the contents as much as possible. But as the atomic number increases, even more neutrons are required. This makes the outer surface of the nucleus more irregularly shaped and causes more angular deviations of the charge inside the nucleus. As these deviations approach and exceed the angular tolerance, the asymmetrical charges inside the nucleus create oscillation, which is intensified by the effect of the outer orbiting electrons. Therefore, even a small particle like a neutron with a moment of momentum casting into the nucleus will cause the barely balanced charges to split and turn the sympathetic vibration into one that is non-sympathetic. The splitting of the two groups releases radiation as the weak force, as Otto Hahn and Fritz Strassman (1902-1980) discovered in their decisive experiment that produced nuclear fission.

---

[5]   The gluing power is still derived from the strong negative charge. If there were a gluon, could it strongly bind two protons together without suddenly discharging all three electrical charges into energy? How could the two protons still attract two electrons, given that their positive charges would have already been cancelled by the gluon?

Nuclear surface tension—derived from the neutral fields of the protons, neutrons, electrons, and the total neutral charge field of the atom—could be called the atomic gravity field. For example, the element fermium has a proton number of 100, but the protons represent only about 40 percent or less of the element's total weight, depending on how many more neutrons are in its isotopes. Having so much neutral energy to maintain the strong force leads to an imbalance at some points, causing the surface tension to exceed the marginal status.

The original gravity field, which derives from the nuclear charges, tends to maintain the atom in a ball-like shape, but inside an element's nucleus, the angular charges are constantly affected by unequal numbers of neutrons, making the angular charges' position in the nucleus asymmetrical and transiting to cause oscillation. Thus, protons have self-contradictory moments of momenta. With more protons and even more corresponding electrons, this self-contradictory status becomes more acute and thus the atom more unbalanced.

Section 4.
Neutral Charge Field

In this section, the "neutral charge field," previously mentioned, refers to the field surrounding the nucleus of the atom, which is quite condensed; while the much less condensed "neutral field" exists outside the atom and in the gravity field, whose charge field is loose. Light is composed of quanta, and a photon is a particle with mass. If a photon's structure breaks upon impact with a certain material, then its shattered content will emit numerous waves consisting of minuscule particles; these are then absorbed by this material because the particles' nature is closer to that of the neutral field, which has the adherence feature. Some of the "-" charge of the prior photon become a portion of an electron, while some of the "+" charge fills in the neutral charge field; when the "+" charge field overflows, a positron is formed, whose charge quantity equals that of the electron's "-" charge field.

As previously mentioned in Part I, the self-spin of the positive charge field will link with that of other positive charge fields. The positron has a similar self-spin and linking tendency. But a single positron has a self-discrepancy in that its two ends will link; therefore, a positron is unstable and will oscillate (Fig. w1). When a positron encounters a negative charge electron (Fig. w2), they will break each other (Fig. w3), and then link to become either photons or neutrinos (Fig. v). When many electrons and positrons meet, they may combine into larger, short-lived particles and then decay into ordinary, smaller particles.

Fig. w

1

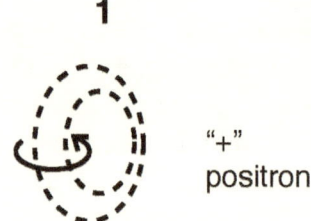

"+"
positron

2

"−"
electron

3

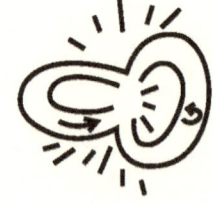

The explosion of many galaxies originates from condensed matter (Part X). When a galaxy explodes, the tremendous outward and inward revolving force disperses into basic particles with a positive charge to create self-spin. This self-spin force, combined with the particles' vibration, indirectly leads the negative charged particles to spin or revolve about the positive ones. This also causes the neutral charge field's neutral fields, both large and small, to tend to form either positive or negative biases, creating a great resonance field, which, in turn, must adjust its resonance frequency into smaller units, such as negative one-third, to match its existence; without matching this frequency, the resonance field will decay into neutrinos and be divided by the neutral charge into increasingly smaller charges.

Therefore, the reason for self-spin or the orbiting of any particles or giant masses is that the positive and negative charges are not totally divisible, which leads to either the repelling or attraction functions. At the same time, the mass of a particle must be related to the nature of its charge and electrical potential. The best guess is that the positive charge is distributed in the center of the particle, while the negative charge is distributed in the particle of the lateral circle.

In physics, the centripetal force results from the total function of the positive charge, while the centrifugal force owes to the total function of the negative charge: These are present in the variety of ordinary motions.

For terrestrial bodies, the lateral turning (repelling) force derives from the negative charge, while the spinning (attraction) force, including gravitational force, derives from the positive charge (for more discussion, see Part X, Section 3). The "+" charge of self-spin and the "-" charge of lateral turning are always perpendicular to each other. Whenever any outer force interferes with this perpendicularity, the natural rectification is for the respective neutral fields to attempt to restore the perpendicular status.

Therefore, the neutral field consists of extremely minute particles—all matter is still energy, congregated into myriad existences of varying durations—and has the potential perpendicularity of self-spin and lateral turning. Thus, it can attach itself to either the "+" field to self-spin or the lateral field to turn. At the same time, it can function as the neutral-charge medium to fill in gaps in the empty space and also absorb less "+" or "-" fragment charge units to form temporary particles. This more or less describes the salient features of the neutral charge field.

The cause of the electron's spin or turning is the proton's self-spin linked to the outside. In addition, the "+" charge of the proton is stronger than the electron's "-" charge. The farther the electron's distance from the nucleus, the faster the electron turns and the easier it is for its attached neutral field to become affected externally (e.g., by friction, impact, or heat). As its neutral field diminishes, the electron becomes lighter and spins super fast. Affected increasingly by the centrifugal force, the still existing electron will flatten into a thin shape in a larger dimension.

In summary, the relationship between the neutral charge field and the "+" and "-" charges is as follows:

- The "+" charge possesses elasticity and plasticity, as well as the ability to loop and link; these combined features cause lineal oscillation. As the construction becomes branic, the already saturated "+" charge loses its ability to loop and link, but retains its elasticity and plasticity; when more than 50 percent of this brane touches another brane of similar curvature, the two will merge into a new brane. This is a major feature of a gravity field.

- The neutral charge field can deliver charges into its own domain; owing to its plasticity, it can mediate between the separated "+" and "-" charges. All radiation carries some of the neutral charge field.

- The "-" charge is lateral but perpendicular to the "+" charge. Enwrapped by its neutral charge field, an electron is not totally free to spin. Once its neutral field is diminished to the maximum extent, the electron spins completely freely without cessation.

- The "+" charge contracts, while the "-" charge expands or stretches; these basic features are evident in an electromagnetic wave propagated in the air. The nature of the neutral charge is to quickly restore the status of a balanced chemical structure. One of the neutral charge quantities is temperature. For example, in a substance that is cooling down, the neutral charge field flows out; conversely, as a substance heats up, the neutral charge field is absorbed.

- The neutral charge field can permeate into the nucleus. The excessive quantity of the neutral charge field will cause the proton and neutron to vibrate and, at the same time, cause the outside revolving electrons to turn less easily and transform into an oscillating energy status. The opposite situation occurs when a low-temperature superconductor is the neutral charge field extracted from the atom. In this status, inside or outside the nucleus, the neutral charge field is the lowest quantity because quarks return to the phase-angle shifting status, causing the proton or neutron to self-spin and motivating the surrounding neutral field to spin in the same direction; when the nuclei spin in the same direction, they will bundle together, and, in the process, push out a great many electrons to the outermost boundary to turn around the central bundle. This is the quantumization of magnetic force, which is super-strong.

- In general, the weak force is the neutral charge field that has permeated out from either the nucleus or between neutrons; it may also cause the alpha particle to be pushed out from the nucleus. In addition to the effect of the

43

weakened nuclear surface tension, the emitted neutron's piercing into the nucleus influences the splitting of the nucleus by its attached neutral charge field, which wedges in to cause the nucleus of the neutral charge field to self-split, and consequently, the nucleus, to split.

Section 5.

Gravitational Branes

This section reiterates discussion on the attraction force of the gravity field. As mentioned previously, the gravity field originates in the center of a body, passing through every atom, layer after layer, tightly combining into concentrated membranes. For a terrestrial body, such as a planet, the outer branes closest to the solid material are the most condensed. A brane naturally contracts, thereby repelling any outside force to interfere at any point on it.

An object's branes are mutually repelling. A freefall object moving toward a planet gains velocity because, as it approaches the planet, its uppermost outer brane, which has the same curvature direction as those of the planet, merge with the planet's brane, and together they contract toward the center of the planet. Underneath the falling object, the opposing curvature of the planet's smaller brane repels the falling object layer by layer, causing the falling object to float. In fact, the falling force is caused by the constant push of branes against each other; as a result, the gravity force is considered rather weak, compared with the color force, color charge, and electromagnetic force.

An electron surrounds its nucleus not only because of neutral-field insulation but also because the curvature of the inner nucleus gravity field's brane is much larger than the opposing curvature of the electron's gravity brane, which turns to gliding. This situation is somewhat analogous to the relationship between the Earth and Moon. The gravity brane of the Moon curves inward toward the center of the Moon; likewise, the Earth's gravity brane curves inward toward the Earth's center. The two bodies' respective outer branes push against each other, causing the smaller body (the Moon), with its opposite curvature, to glide into orbiting the larger one (the Earth). However, the two bodies mutually create much larger, outermost branes, which hold them together.

Therefore, gravity branes—from atoms and molecules to material objects—merge as long as their respective curvature directions are the same. If opposing, they repel each other.

It is the unbalanced charges in the universe that always cause motion.

PART VII

Gravitational Force and Inertia

Section 1.
Gravity and Inertia: How They Differ

On August 28, 1918, in his letter to A. Bestmeyer (1875-1954), Albert Einstein expressed that his general theory of relativity was [based on] the equivalence between inertial and gravitational mass (i.e., the equivalence principle).

From the smallest particles (including photons and neutrinos) to the most grandeur—all gravity is connected into a grand gravity field as long as their curvature direction is about the same. One can use a separating line of 180 degrees.

According to the general relativity principle, everything in the gravitational field has the same velocity. But this book disagrees with the concept of an evenly distributed gravitational force field (as mentioned in Part IV).

In the gravitational field, not all matter has the same acceleration speed; lighter-mass objects accelerate faster than those with heavier mass. To illustrate, one can imagine that initiating the acceleration speed of a vehicle or rocket would be faster when emptied (zero-loaded) versus fully loaded. Or one can imagine setting a single paper clip alongside a bundled group of 10 paper clips of equal size the same distance from a large magnet, which is pushed toward them. The single paper clip will move toward the magnet at a faster speed than will the bundled group. Or if two paper clips of equal size are placed at different distances from the magnet, the initial speed of the one closest to the magnet will be faster.

From these experiments, one understands that:

- A matter's attraction field does not exist relatively; that is, an atom or molecule exists naturally, meaning that no other matter's relative existence can increase or decrease the force of its attraction field. The original matter's attraction field exists, regardless of whether other relative matter exists nearby. Otherwise, all of the molecules in a terrestrial body would disintegrate. But, in fact, the gravity field of every particle always exists independent of the terrestrial body's existence.

- In a terrestrial's body's gravitational field, all matter, owing to their various masses and distances to the center of the terrestrial body, have different velocities.

- The mass and density of the terrestrial body will differ in places and thus the strength of gravity will differ.[6]

- When a terrestrial body self-spins and its surrounding gravity field accelerates into inertia motion, the falling object's velocity will not be exactly the same at every point. On Earth, for example, the velocities at the two polarities are faster than the same-height velocity at the equator.

- If a terrestrial body contains some moving matter, such as water, then the gravitational force above this matter will be weaker because the vector of every molecule in the moving part differs and thus the gravity force cannot totally combine at any time. Since the Earth's ocean water

---

[6] For example, in some places on Earth, cars roll uphill with the engine off because the underground mass and density at the top of the hill are heavier than those at the foot of the hill.

is always unstable, the gravity force above the ocean (same height as to the ground) is weaker.

The Earth's gravity branes lack a fixed curvature because of the varying underground masses and densities, differences in molecular motion, and varying magnetic-line strength.

The vectors inside the nucleus explain why two elements with high atomic numbers that impact in a high speed form an element with an even higher atomic number.

Gravity is purely a part of any mass. But once a mass is in motion—whether acceleration, deceleration, spinning, orbiting, or uniform motion—it is in inertia. In the universe, matter in a pure gravitational status could seldom be found, suggesting that matter is not in inertia. In space, the curvature of the gravity field is always changing as it is constantly affected by the motion of other bodies' inertia or gravity fields.

Section 2.

Gravity and Inertia: About Curvature

Gravity differs from inertia in that it exists in and of itself. By contrast, inertia refers to states of rest and motion: At rest, inertia directly represents the status of gravity, while in motion, it represents the change in gravity's status—that is, from an almost perfectly rounded form into an elliptical or other shape.

Therefore, the curvature gravity causes matter to have motion, and the matter's own motion changes the curvature of the gravitational field, which is the opposite of the simplistic notion that fixed curvature can affect motion.

The Earth and Moon can be taken as an example. They have their respective gravity fields, which are perfectly round. But because they are in motion, having both self-spin and orbiting of each other, their gravity fields become flattened and elliptical. Thus, gravity's curvature differs totally from that of inertia.

Section 3.

Gravity and Inertia: Layers

Extending the discussion in Part IV, the curved layers of gravitational branes (membranes) and natural curved layers can be blocked, and a body's self-spin can flatten its many ball-shaped gravitational layers into flattened inertial branes. As Fig. x1 shows, the outermost branes are laterally expanding and perpendicular to their imaginary axles. If the spin speed is faster, the inner branes will change less but will be condensed, while the lateral ones will expand into a flattened plate, which will block the other large gravity-curvature branes (Fig. x2).

This situation is akin to suspending a superconductor in the air by placing a magnet above it, in which case the magnetism is lineal between electrical charge fields. But in this case, the gravity is a brane field. In Fig. x2, the gravity field of matter m, owing to its self-spin, changes into a lateral inertial field. Since the gravity field is a repelling, rather than an attracting, force, an interterrestrial body, such as a planet (matter E), simply uses its outermost gravitational branes to enwrap other nearby objects. Though the result appears as mutual attraction, it is, in fact, mutual enwrapping.

The high spin speed of the smaller matter (matter m, Fig. x2) flattens its curvature; and the flattening gravity brane repels to restore its natural curvature, which pushes matter m. The more matter m extends its lateral dimension, the more the even stronger force of the underneath gravity brane (matter E, Fig. x2) will push back.

To reiterate, since gravity originates in the atom, why is it far less powerful than the strong, weak, and electromagnetic forces. The answer is twofold. First, the original force, with its $4/3\ \pi c^3$, stretches out completely, and this volume is quite large. Naturally, it will degrade any point in its domain's intensity, making gravity and the other three forces minute by comparison. To measure the strength intensity at any point, one should use $4/3\ \pi r^3$. The increased value of r causes a reduction in the gravity force. Therefore, extension of the gravity field is not without limit.

Secondly, because the gravity field is one of repulsion, not attraction, each of its branes has the enwrapping and contracting nature. Thus, any two consecutive branes repel each other. And since two repelling forces cancel each other, that causes the curvature to differ and renders the remainder of the force relatively weaker.

Furthermore, if matter m has a larger quantity and spins faster, it will float above matter E's even stronger force. Even more apparent, it will block matter E's gravity field and enforce its own smaller, but independent, gravity field. If its lineal magnetic field causes rapid self-spin, that can isolate the functioning of its extraterrestrial gravity field.

Fig. x

1

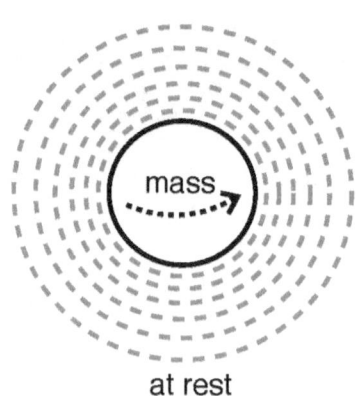

at rest

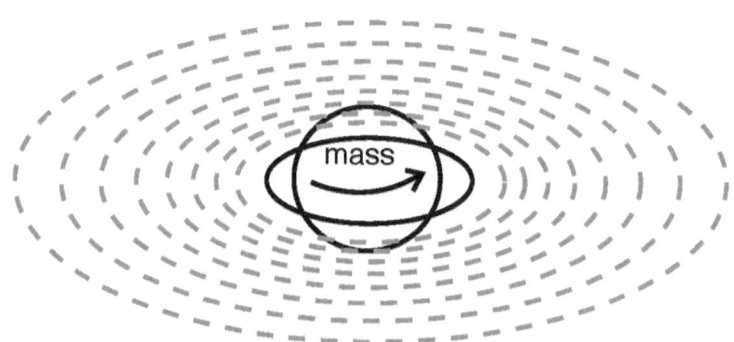

During fast spinning, the mass itself,
as well as the branes of gravity,
flattens. Branes of gravity are
a part of mass, both of which
are affected by centrifugal force.

Fig. x

2

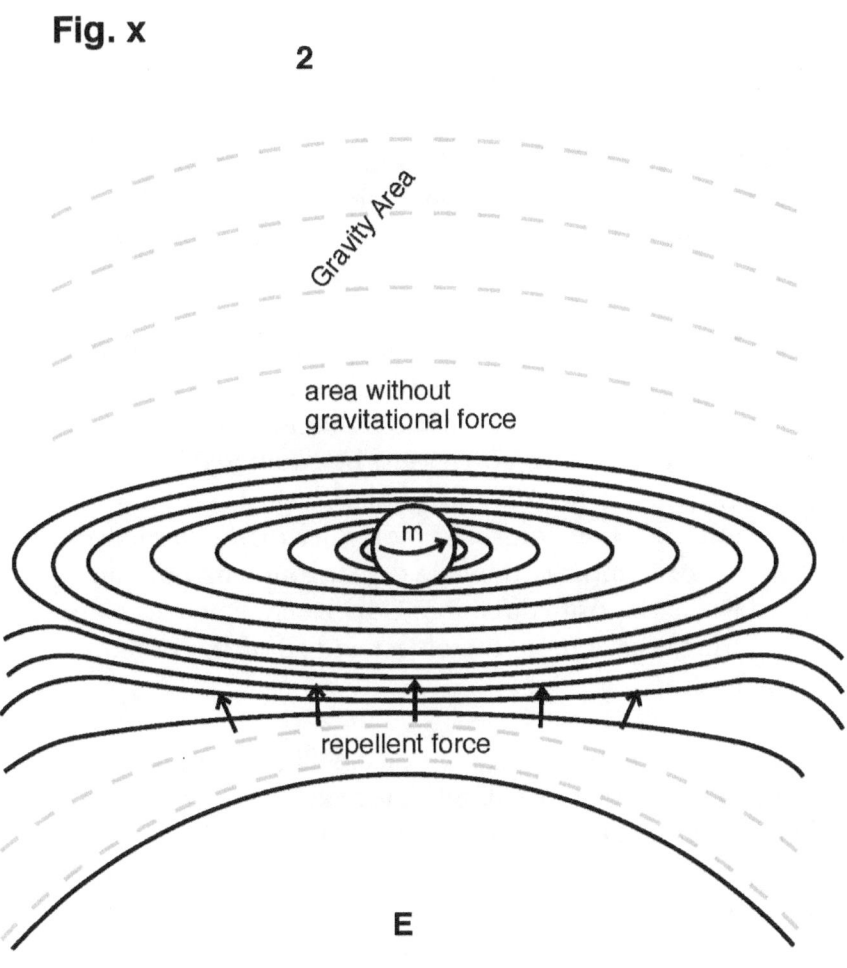

Gravity Area

area without
gravitational force

m

repellent force

E

A completely flattened, artificial body and dimension passing over a planet's curvature can block the planet's surrounding magnetic field. In such a case, the flattened artificial body of whatever volume will float above the planet while hardly moving, and follow the planet's own gravity to spin or orbit.

Section 4.

Hidden Energy of Uniform Motion

Part IV, Section 2, raised the question of the hidden momenta of a chamber in uniform motion. Overall, the inertia field of an accelerating object will resume into the G-field once acceleration ceases, but the momenta still exist.

Part V, Section 4, concluded that an object in motion does not contract toward the moving direction but instead is dragged in the opposite direction of the motion; that is, an equal anti-force exists: A forward pushing force must be countered by a dragging or pulling force behind. When an object is pushed during acceleration, only the atoms and molecules of the immediate area pushed will be compressed, while the surrounding ones will be dragged in the opposite direction. Conversely, when an object is pulled during acceleration, the object and all of its contents will be stretched out.

Once acceleration stops, the object will be in uniform motion, and its in-motion gravity will be its inertia. This inertia's momenta will be maintained in all of the object's atoms and molecules (i.e., they will follow the features of the object's nuclear charge field). But this molded shape does not represent the moving object's entire vector. Simultaneously, the construction of the object's atoms and molecules has the complex positive- and negative-charge forces. The construction's charges have their angular momenta, and there are negative-charge angular momenta between the molecules. In addition, the object's material has its chemical structure and thus its unique resilience ability.

When an object is in uniform motion, the closest branes of the inertia are on the vector's molded status. To the external domain, these gradually become gravity, which is the status of inertia at rest. But in gradual acceleration to the speed of light, an object's gravity or inertia field is not limited to the speed of light and may exceed it, which is the status of different universes having different layers.

The above explanation may account for some of uniform motion's hidden energy. But the major energy of the momenta is still in the uniform motion's solid matter and its surrounding inertia. As Fig. o2 illustrates, the two riders in separate chambers moving in opposing directions at a high rate of speed will feel they are at rest in space, but an outside observer can see what is really happening. Thus, the energy of the momenta is directly on the moving object's inertia. In daily life, we can see that momentum is with the moving object, which carries along its inertia.

As explained above, the vector of an object's uniform motion is in all of the object's atoms and molecules and their mutual inertia fields. Then what is the end of those vectors of the uniform-motion object? The object may be blocked by an anti-force, which produces new vectors or may crash, with its broken pieces generating new vectors. But if there is nothing to stop it, the object will carry its vector to enter the gravity branes of the extraterrestrial gravity, making the momentum of those vectors transform into surrounding inertia or enter the more condensed gravity branes. That will cause them to change into either left- or right-spin energy momentas; in this situation in the particle universes, those types of transforming force change into orbiting or self-spin energy.

Finally, it should be mentioned that, once acceleration ceases, the object's own inertia field will return to the gravity shape, pushing the object's solid body backward (opposite to the direction of the motion); that is, owing to the repelling force, the object's own gravity field will push the object to the center. The object's solid body will retract a bit from the highest speed, meaning that the uniform-motion speed of the whole object, including its inertia field, will be slightly slower than that of final acceleration.

This explains why we can observe celestial bodies either orbiting or spinning. Initially, many of them were in uniform motion, including centrifugal tangential line motions.

Section 5.

On Inertial System

Only inertia at rest is the normal gravity status. In the states of self-spin, orbiting, uniform motion, and simultaneous multi-inertial motion, the structural angles of the constituent atoms and molecules differ from those of the angular momentum of pure gravity. This will affect molecular chemical functioning and thus biological functioning over the longer term. Because of such differences in inertia and gravity, the bio-structures of creatures living on other planets will vary.

Section 6.
Conclusion

An object's inertial system—comprising inertia at rest and the inertia of self-spin, orbiting, spiral motion, uniform motion, as well as the complexity of combined inertial motion—accumulates in the object's electrical charge fields. And in motion, this inertial system combines with the object's external inertia field.

PART VIII

Moon-Earth Relationship

Section 1.
Gravitational Functions

Since gravity is a repellent field, why do the ocean tides move upward (convex curve) instead of downward (concave curve) as the Moon periodically moves closer to the Earth's surface?

When the Moon faces the Earth's ocean, the mutually repellent force between the two bodies causes the formation of flattened branes (Fig. y1). At this time, comparatively more of the Earth's surface matter will receive the dimension of the upper branes, versus the lower ones, causing the more steeply curved gravity field of the Earth's crust or mantle to be repelled out; that is, the Earth's surface matter is pushed out by the repellent force underneath. Thus, what appears as attraction by the Moon is, in reality, repulsion from the Earth.

Section 2.
The Moon's Substantial Effects

The speed of a meteor moving toward the Moon is accelerated centrifugally by the Earth's self-spin; attraction by the Moon's gravity field, without the encumbrance of air, also accounts for a small portion of this accelerated speed (Fig. y2).

Fig. y

1

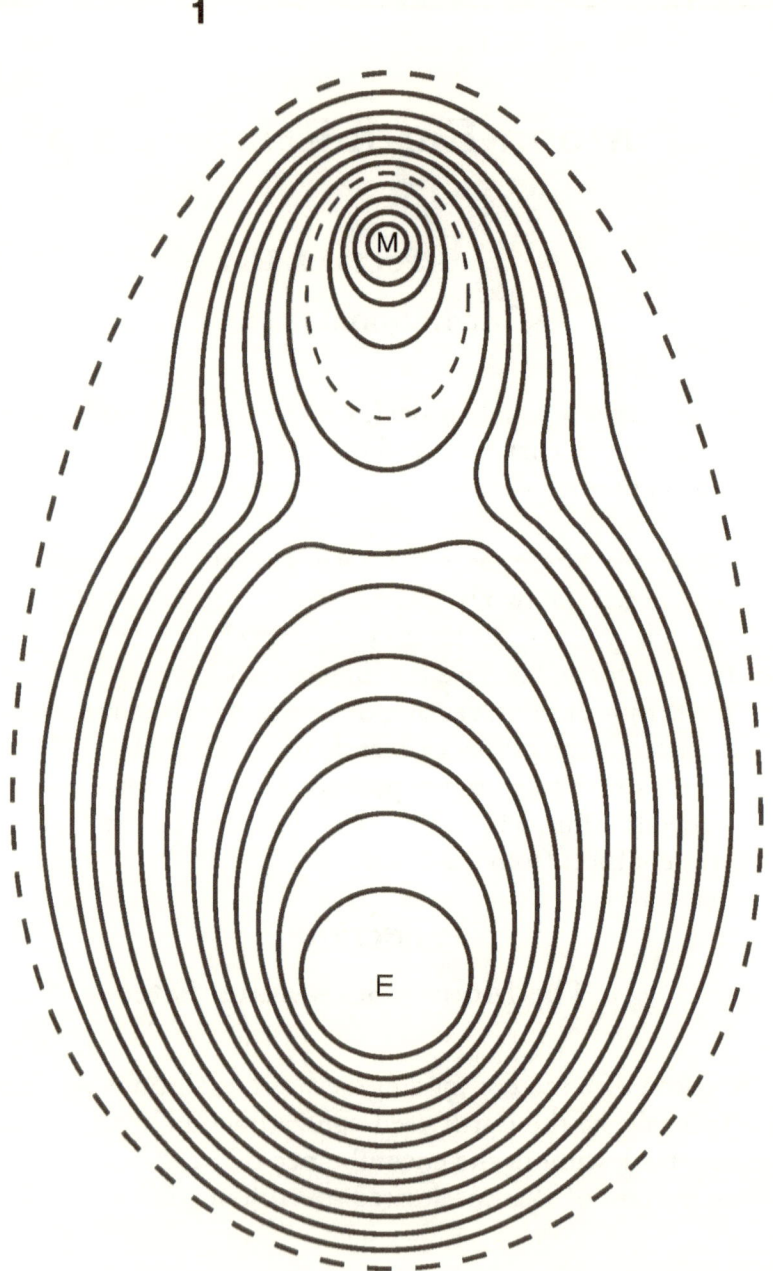

Fig. y

2

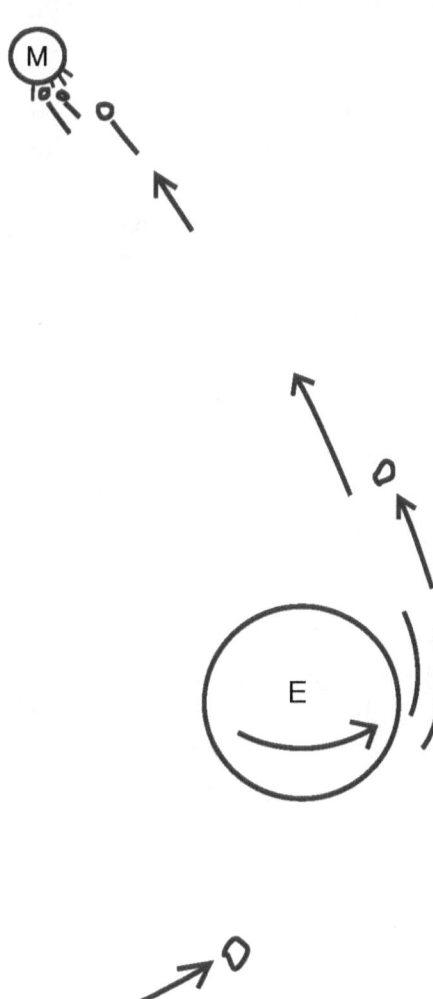

When the meteor's dense matter impacts the Moon's surface, the two compressed materials will squeeze out each other's neutral charge fields, and the atomic numbers of the elements in the newly formed material will nearly double.

Since a star's fuel from hydrogen, plus helium, starts the fusion process, which stops at iron, with 26 protons, the atomic number of the elements produced cannot exceed that of cobalt or nickel. Therefore, it can be expected that the high-speed impact of two dense celestial bodies will produce many elements with higher atomic numbers.

When a meteor hits the Moon, the volume of the latter is not enough to produce elements with higher atomic numbers. Even though debris from spatial remnants can be discovered on the Moon's surface, finding elements with higher atomic numbers would probably require traveling to planets with a much greater volume than that of our Earth's mass.

In the known universe, as on Earth, one could similarly expect to find many radiational, high-atomic elements produced from the high-speed impact of two high-density materials. Perhaps on Earth, using an accelerator, high-atomic elements with more than 110 protons could be produced by compressing heavy metals together.

On the Moon's surface, some of the Sun's light is reflected by metals with high-atomic numbers, while another portion of its energy is absorbed. But reflection and refraction produce spectrums of varying proportions; thus, the effect of moonlight differs from that of sunlight on certain life forms.[7]

---

[7]   For example, a certain type of jasmine plant blooms only when moonlight has been cast on it; however, its blossom is purple rather than the white color commonly associated with the jasmine plant. The various elements on the Moon's surface may cause different spectrum distributions, to which this particular jasmine plant responds, according to its own spectral selectivity and sensitivity. From its leaves, the plant can absorb this

The counterforce effect of the constant bombardment of meteors onto the Moon's surface from the Earth's acceleration will more or less push the Moon away from the Earth.

Section 3.

Implications of the Moon-Earth Relationship

Not only does the Moon affect the Earth's ocean surface. When it is directly above the planet, everything—from the blood in human beings' heads to the land itself—is pushed upward in the Moon's direction. Therefore, when the Earth is facing the Moon, the density of the soil or land will loosen and even the planet's magma will expand a bit. As the Moon recedes, the soil or land will then fall downward (but not as low or dense as before), and the magma will fill in the minute gaps (its density will return to normal). This continuous up-and-down phenomenon affects the Earth geologically.

The Moon may be older than the Earth, but its volume is lighter. Its volcanic movement stopped long ago, yet the Earth still enjoys much seismic activity. Thus, the Moon is quiet within, while the Earth is loud; the Moon cold, but the Earth hot.

If one were to suppose the Earth is cooling, then, like the Moon, it eventually would have no earthquakes, volcanic eruptions, or other geologic events. If true, then all of the planet's mountains would disintegrate into rocks and its land into sand and dust, subsequently flown or blown into the ocean. In such a

---

information into its root, and then search for an element with a similar spectrum frequency. Some of the Earth's soil may contain pyrolusite (MnO_2), along with some magnetic manganese (Mn) (atomic number 25) and possibly a bit of plutonium ions (Pu^{+3}), all of which are purple in color. This would explain the purple color of the plant blossom. This example suggests that some special elements that absorb sunlight or artificial white light could be used to reflect light onto certain plants to create new varieties.

situation, no north- or south-polarity ice caps would exist because nothing would impede the flow of hot and cold air currents. In short, the entire planet would become covered by water, and all animals and plants would devolve into marine life: A horrific scenario to ponder.

We know that the mountains created from geological shifts are a necessity for sustaining land creatures; thus, we must bear the effects of major seismic events. Since the Moon is not hollow but is rather a solid satellite, its gravity field can affect the Earth's up-and-down movement.

Because the Moon's mass is too small to catch the Sun's major energy-neutrinos, these continuously pierce through it, explaining why the once hot lava within the Moon long ago cooled into solid material. In the case of the Earth, most of the Sun's energy-neutrinos pierce through it; however, unlike the Moon, the Earth can catch some of these neutrinos, owing to its larger volume and density and the effect of its gravity curvature on neutrinos. Those curved at the edge of the Earth's crust can be absorbed into heat energy: the source of the lava mantle that keeps the below-surface Earth bubbling with activity.

Sole dependence on the Sun's energy would not suffice to produce the geological movements needed to form mountains. That alone would only produce the Earth's heat and cause massive steam to surround the planet. Rather, the massive movements responsible for mountain formations and crust changes require cycled motive force, which naturally derives from the Moon's orbiting of the Earth.

The orbiting Moon is not only responsible for the Earth's ocean tides, but additionally the "tides" of our planet's land and mantle, including the lava. The effect of the continuous up-and-down movement on the land, unlike the effect on water, means that the gap beneath the raised land will be filled by mantle, meaning that the land cannot fall back completely. Over long periods of time, this repeated action gradually accumulates energy and position, finally causing the large movements that form mountains and continental crust lumps. To a lesser extent, the up-and-down movement also affects movement on or below the Earth's surface.[8] Even during the Earth's self-spin, the ocean water and its currents affect the continental land lumps.

In conclusion, the Moon, as the Earth's satellite, is indispensable for maintaining life on the land surface. To sustain all land plants and animals, as well as human civilization, the Earth's surface cycle is thus inevitable.

---

[8] Of course, the Sun's gravity also affects this movement.

PART IX

Neutral Charge Field: Further Thoughts

Section 1.
Basic Questions

This chapter focuses on applications of the neutral charge field and particle theories proposed thus far. Sole dependence on the positive and negative charge is insufficient to explain material existence or to answer a host of questions pertaining to natural phenomena. For example, why is the explosion of two materials that collide hexagonal in shape? Why do snowflakes always have six angles? If three pairs of polarity make a shape, why is each angle between these pairs 60 degrees? And why do they form into hexagonal planes instead of a cubic hexagon? How it possible that three pairs of magnetic polarity that intersect at a common point extend out in straight lines without interference?

In the context of this book, we can only guess at the answers because, in either a proton or neutron, the basic electrical charge is one-third. The angular deviation of their charge is equally indivisible; therefore, they always have a repelling force that causes them to self-spin and induce electrons to spin or orbit. The nature of positive charges is to form a straight line; owing to their one-third feature, after passing the point of intersection, they become six times 60 degrees to form a complete circle. But why is the circle a two-dimensional plane instead of a three-dimensional sphere? The answer is that the positive charge, with the negative charge, forms a perpendicular angle, which deviates eccentrically.

Under high temperature, pressure, or electrical charge, they are forced into angular dependence of lateral scattering. In this situation, the vertical and horizontal charges change position, resulting in a six-angle shape. Because the instantaneous negative and positive charge-field phases are not perpendicular to each other, the difference in vibrations causes their energy to scatter. Most of the considerable energy squeezed out consists of a di-electric, neutral charge field and electrons. Under frozen material, the neutral charge field leaks out, and the electrons' spin speed, range, and dimension increase, rendering electrons and protons; the proximity of their mutual attraction force causes the matter to tend to deviate into two dimensions (e.g., snowflakes).

With regard to non-interference of three pairs of magnetic polarity, the extra neutral charge field surrounding the point of intersecting lines causes the point to become multi-directional, thereby allowing the three differently-angled polarity lines to pass through unimpeded.

Summing up, in the universes, the three-dimensional material is quite stable, due mainly to the neutral charge field.

Section 2.

General Conclusion

In general, the gravity field structure has three charges: (i) the positive charge, which radiates out from the center of any matter into a lineal structure; (ii) the negative charge, which surrounds and expands into the positive lineal structure; and (iii) the neutral charge field, which links them both into a curved phase and, in turn, a brane structure. Every brane's phase is connected to the gravity field of any matter's atoms and molecules; because of the deep penetration of the gravity field, every atom of a matter possesses the brane contracting function force. But the high-speed self-spin of a smaller flying body can block the much greater gravity field of a larger neighboring body (e.g., a planet, such as our Earth) in order to form the smaller body's independent gravity field. At the same time, owing to high-spin flattening, the independent gravity field of the smaller flying body has enough repulsion to force either air particles or water molecules aside, meaning that the spinning body's movement is not affected by the friction of air or water molecules.

Section 3.

Neutral Charge Field and Non-crystal Metals

The neutral charge field can be described as a di-electric substance that functions as an atomic filler. For this reason, when hydrogen lacks one or two neutrons to compensate for its single proton's positive charge, the proton requires a large quantity of neutral charge field to fill in between the nucleus and the electron. In the nucleus of uranium, the quantity of the neutral charge-field filler is relatively smaller because the number of neutrons is greater than the number of protons. Thus, outside the proton-neutron binding, the quantity of neutral charge-field filler is relatively less. Therefore, the process of uranium fission releases less neutral charge field and thus less energy than does hydrogen fusion.

Non-crystal or amorphous metals do not occur in nature.[9] Rather, they are produced artificially by high-temperature melting of an alloy, followed by its quick cooling (at a rate of 10,000-1,000,000 °C per second). But to prevent the amorphous metal from reverting to its crystalline structure at higher temperatures, laboratory research discovered that a minimum of three metallic elements are required for the alloy.

The advantages of amorphous metals, compared to ordinary crystal metals and their alloys, are numerous. For example,

---

[9]   Laboratory research on amorphous metals began in about 1940, when it was discovered by chance that the Ni-P alloy was non-crystal. Twenty years later, the Au-Si alloy was found to have an eutectic-reaction structure.

amorphous metals are a million times less corrosive and have about 10 times more magnetic permeability. They have lower coercivity, a better soft magnetism property, and higher electric resistance. Because they are without grain boundary, there is no segregation, which causes corrosion. In addition, they reduce magnetic loss (i.e., minimize eddy-current loss), are hard, and are neutrons-impact tolerant. They also have a useful catalytic function, high hydrogen absorption, and a strong radiation feature.

But the structure of amorphous metals is not without its major defects. For example, in a higher-temperature energy environment, it tends to lose its non-crystal feature and revert to its crystalline form. Various methods have been developed to make amorphous metals (e.g., quenching, electroless-plated coatings, plasma spattering, and ion implantation); however, the resulting products have been of thin volume (e.g., metallic film, belts, thread, or small granules). A greater need is to learn how to produce relatively larger-volume amorphous metals.

Section 4.
Crystal Structure of Metals

Because of chemical binding, non-crystal alloys run the risk of reverting to crystal metals when high heat is applied. To prevent the risk of such occurrence, this author suggests that using physical, instead of chemical, binding may render these non-crystal alloys strongly resistant to heat, corrosion, and breakage. For metals, this book assumes that the strong force between protons and neutrons possesses the electrical-charge twisting angles, meaning that each element may have more positive charges of the angular focusing point (e.g., angle of convergence) or, conversely, more negative charges of the angle of repose.

The crystalline structure of natural metals may be regular or irregular. The three major regular crystal structures are the (i) body-centered cubic (bcc), (ii) face-centered cubic (fcc), and (iii) hexagonal closest-packed (hcp) (Fig. z). Of the three, the hexagonal pillar crystal (iii) appears to be the most common. As illustrated in Fig. z1, the bcc crystal contains six square-based pyramids at the angle focusing point (point o); metals that typify this shape include β-Ti, Cr, α-Fe, δ-Fe, Mo, and γ-U. Likewise, the fcc crystal has six square-based pyramids, except that their respective vertexes lie on the same plane (Fig. z2). For this reason, the center of this cubic crystal apparently lies in cubic space; examples of fcc-structured metals include Al, γ-Fe, β-Co, Ni, Cu, Pt, Ag, and Au.

Fig. z

crystal lattice

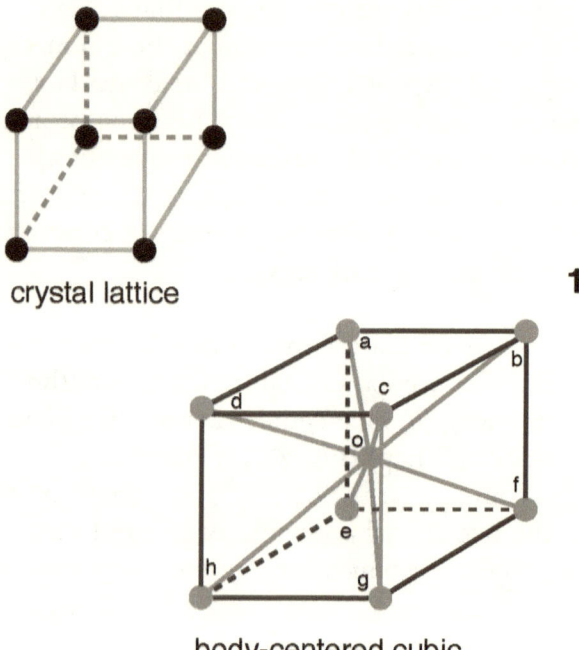

1

body-centered cubic

2

face-centered cubic

Fig. z

3

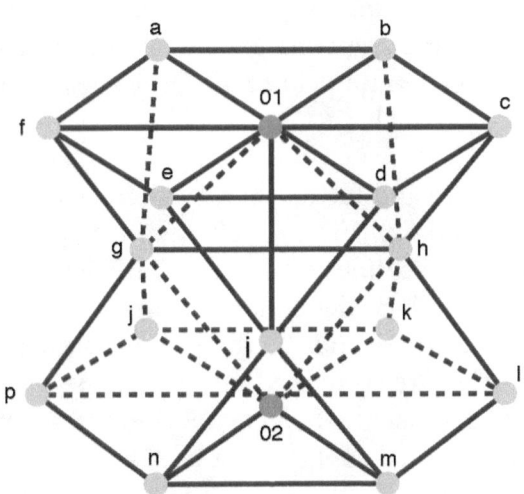

hexagonal closest-packed

The hcp-structured crystal has only two vertexes, as illustrated in Fig. z3 (points o1 and o2). The top half and bottom half have three square-based pyramids each, for a total of six. But at the center of this crystal stands a three-dimensional diamond, a combination of two 60-degree angle, three-dimensional triangles. Therefore, the three angles between the bases of the two center triangles are each 120 degrees. But neither their shape nor the six sides are solid, given that the distance between the top and bottom six atoms is too far.

It thus follows that, in nature, the square-based pyramid structure becomes essential. As suggested above, the six square-based pyramids of the bcc-structured crystal have only one vertex (point o) (Fig. z1). By contrast, each square-based pyramid of the fcc-structured crystal has its own vertex, for a total of six (Fig. z2). Both the bcc and fcc crystals possess convergence into angular focusings at point o, which compress their phase angles into bunching emissions in six directions. These two types of crystals, containing either 45- or 90-degree angles, may be categorized as such.

Unlike bcc and fcc crystals, the hcp structure has two center points (o1 and o2) (Fig. z3). The hcp can have six square-based pyramids, whose inner angles are all 60 degrees, each of which has a 30-degree angular dependence. Thus, the total repulsive potential angle is 60 degrees, and the three-dimensional triangles total eight, two of which are combined into the shape of a three-dimensional diamond. Because the hcp crystal is in the form of matter, the repulsive potential energy must be less than the convergent potential energy. Furthermore, the total volume of the six square-based pyramids is twice that of the lattice formed by the eight three-dimensional triangles. Thus, the hcp energy structure is based on the six square-based pyramids, and the crystal element type is categorized as 60 degrees plus 30 degrees. In summary, all three crystal structures have their energy-of-convergence angles, all of which are formed by six square-based pyramids. Could this be the natural form of energy convergence?

Section 5.

Amorphous Possibility

In this section, it is proposed that amorphous metal does not exhibit the ordinary chemical binding of non-crystal metals. It is commonly known that more than three types of elements can bind at higher temperatures to attain the more stable structure of an alloy that cannot easily revert to its natural crystal form. In addition to ordinary chemical binding (i.e., the binding of "+" and "-" ions), alloys can be produced from an inter-metallic compound, which is much stronger. A good example is the inter-metallic compound, $CuAl_2$.

A third form of binding, introduced by this author, attempts to push beyond the non-crystal metals technique to create super-strong binding. It requires at least three elements, uses the process of rapid cooling or quenching to extremely low temperatures, and involves combining the neutral charge field's metallic elements.

Deciding which elements to combine requires examining the nuclear structure of all elements. A nucleus in which each proton carries one or two neutrons will cause different angular potentials of the nuclear-binding charge.

Examining the elements and their isotopes is based on the author's theory concerning the neutral charge field:

a) Because the neutral charge field (though minute) carries the self-contradictory force, it is always oscillating matter or motion energy.

b) Owing to this feature, the neutral charge field can adhere to both positive- and negative-charged particles.

c) Because of its strong adhesive feature, the neutral charge field can maintain the pattern of the trace of the moving particle, as in the case of polarized light (previously mentioned in Part II).

d) So-called heat energy is formed by an excessive neutral charge field that has permeated matter to cause vibration or oscillation.

e) The more neutral charge there is between two elements, the easier it is to split them using external, high-heat energy.

f) A thinner neutral charge field between two elements may draw them closer for their conjugate charges; the positive charge with the neutral charge between protons and neutrons—that is, binding power—together with another element's positive and neutral charge, become a mutually combining configuration.

Based on the above theory, it is possible to examine numerous elements and their isotopes to discover the three-dimensional angular spread and tolerance that will yield a mortise and tennon to join certain elements.

From the discussion in Section 4, it seems that, for any particular element, not less than three or four others can be found to dovetail it, thus making the multi-element, three-dimensional structure of non-crystal alloys possible.

The nuclei of two elements can be pulled closer by using one of the element's nuclear angle of charges, compensated by the other element's angle of charges, to quoin in to form a binding structure, utilizing a mutual neutron (instead of the mutual electron used in chemical binding). In addition, the number of elements in this non-crystal alloy is the same in order to continue the molecular linkage. Because the degree of the angle of charges differs for every element and its isotopes, the non-crystal alloy will form a non-flattened structure, rendering it extremely strong and hard. Even extremely high temperatures cannot cause it to revert to its former crystal structure because no extra oscillating space for the neutral charge field lies between this alloy's elements. Moreover, there is no extra space between these elements for the electrons to move freely; thus, electro-resistance is quite high.

For this type of alloy, using liquid helium, along with quick quenching of the temperature, to squeeze out the normal neutral charge field in the nucleus might not suffice to form a perfectly even distribution of the elements and perfect homogeneity of their molecules. Owing to the extremely fast speed of the cooling down process, a high magnetic field and high-voltage electrical field may possibly be needed to help the nuclei dovetail in the most efficient way.

Regardless of whether one uses a blend of metallic powder under high compression or a homogeneous metallic liquid, the rolled-sheet process will require that a high positive voltage be connected to the roller; once the temperature is lowered to the required level, the high-temperature process should be repeated so that the high positive-charge rollers can extract even more electrons. Using an instantaneous quenching method, this super-metal alloy can be treated with liquid neon, instead of liquid hydrogen or liquid helium, to extract even more electrons from the alloy.

If this super non-crystal alloy can be made, its structure will be super strong, and both its heat-expansion and cooling-contraction ratios will be quite small. Its high-resistance feature will cause the electrons to turn into multi-directional refraction and radiation energy, including homogeneously emitted photons, when a current is connected to it. Because this alloy does not contain enough neutral charge field to support heat oscillation, it is quite energy efficient; thus, such material can be used as new sources of light.

If this light source is very strong and condensed, it can isolate gravity branes because the light emitted from the alloy travels at the same speed as that of the connected gravity branes. To free the super-metal alloy from the bond of the gravity branes, liquid electricity should be used to supply the strong energy needed due to the loss of the quanta.

The above-mentioned method utilizes the several elements' nuclear-charge symmetric angles to force the other various elements' nuclei to draw closer and, in turn, push out much of the neutral charge field, limit the electrons' domain of motion, and create high permeability and the low coercive force. Thus, the product will have tremendous heat resistance, a super-strong structure, and high radiation.

Section 6.

On Electron Motion

In non-conductive material, the outermost electron layers do not orbit since the limited domain of their neutral charge field makes them less active. By contrast, electrons in conductive material have more room to orbit when voltage applied at both ends causes an electron at one end to be pushed in, while an electron at the other end is pulled out. Electrons shift from one atom to another in a lineal chain reaction (i.e., by taking the shortest route). That is, the atom captures a fed-in electron, which sets in motion an electron replacement reaction, with each successive atom's electron shifting to the track of its adjacent atom's outermost layer. Thus, the electron that finally exits the conductive material is not the same one that was fed in. In short, a current of electrons does not travel through conductive material; what occurs is electron replacement.

The "tunnel effect" observed in some electronic devices is a stream of quantum particles that encounters a barrier. Some electrons bounce back or reflect, meaning that their self-spin counters the self-spin of the barrier's electrons. But other electrons may pass straight through the barrier; these gain their enhanced force from the barrier's electron spin, which is in the same direction, thereby enabling these electrons to exit at a faster speed.

As the quantum particles flow to meet the barrier, the energy at the barrier's meeting domain will disperse to a certain extent, depending on the intensity. If the particle stream is in a constant speed and about the same quantity or in a rhythm, the barrier will rhythmically respond to allow more quantum particles to penetrate through; otherwise, the barrier may make their passage more difficult, which explains the dwindling effect of the quantum particles flow in the tunnel.

In the case of an electron leaping from one level to another, this phenomenon is related to the neutral charge field rather than the electron's own "decision."

Since atomic elements vary in their number of protons, neutrons, electrons, and isotopes, energy fed into an atom may cause some electrons to orbit into oscillating or cause some to orbit on an outer track. At the same time, the new pack of neutral charge field causes the original atomic neutral charge field to readjust by resonancing in rhythm. In some cases, the orbiting electron experiences the distance variations since the nucleus may not be a perfectly smooth-surfaced sphere. Until the oscillation of the resonancing rhythm has achieved a perfect cycle, the extra energy will be squeezed out. The cavity left may permit the electron to fall back to its original track. But the environmental energy level of neighboring atoms is also a contributing factor. Thus, the electron is constantly affected by alternating stationary and changing states, which cause it to oscillate or fall back: The uncertainty of an electron's motion is too complex to be calculated simultaneously; thus, environmental factors, not the electron's "own decision," cause it to fall back to its original track.

If elements could be artificially compressed closer to exceed more than one layer of the electrons' track, the elements' nuclei would be close enough to share neutrons (similar to the sharing of electrons in chemical binding). In neutron binding, some electrons could be squeezed out and others could be framed. At a minimum, the outermost layer would no longer orbit, and the total charge of the connected atoms would be in a "+" ionic status. Applying a high-voltage electric current to a material with such an atomic-connections structure would not conduct electricity, and the electrons would be transformed into large-volume quanta radiations.

With the occurrence of such a phenomenon, as mentioned in Section 5, a piece of this non-conductive super alloy could be used to replace some ordinary light bulbs or light emitting diodes (LEDs) since the alloy's material would have become quite energy-efficient.

Section 7.
Closing Remarks

As previously indicated (Section 4), the structural system of most atomic crystals is isometric (some angles are 90 degrees; others are 45 degrees). Other atomic crystals have 60-degree angles, with a 30-degree angular dependence. Since the magnetic lineal structure of the electron's motion is perpendicular to the positive axle, which is on the flattened distribution coordinates, there are four upper and four lower vertical, 90-degree quadrants. When a group of electrons is emitted into this space, every electron group will be attracted by the positive-charged antenna. Therefore, the antenna can be designed into 90- and 45-degree or 60- and 30-degree structures. But if the electron group were to be emitted from a pure crystal metal, this group's structure would likely be the strongest.

In general, matter is structured according to the force of the positive-charge structure. Negative charges repel each other, but a group of negative charges can be pulled together because they are surrounded by and linked to the neutral charge field. Therefore, electrical waves in the air generally have a negative charge. These can be called "electro-wave" since they lack the positive charge. Only when attracted by the positive charge do they become electromagnetic force. Thus, if electromagnetic waves (containing the positive charge) are in the air,[10] an antenna with the positive charge cannot receive them; instead, it will repel them.

---

[10]   To qualify as magnetic, a force or a wave must contain both positive and negative charges (see Part I for a detailed explanation).

Since the same atomic elements exist throughout the universe, how would one go about creating a super alloy with extremely strong and versatile features? To meet these high-performance requirements, this book suggests testing five lightweight elements: magnesium, aluminum, calcium, titanium, and barium (with their equal counts of atoms in every element).

PART X

Expectations

B ASED ON THE SYNTHESIS OF DISCOVERIES in this book and
those presented in the previous volume (Parts I-IV), this part
discusses their implications for re-evaluating the structure and
future understanding of physical nature and phenomena of particle
universes.

Section 1.
Neutral Charge

This book and its predecessor volume have often mentioned the
neutral charge. The conventional thinking is that the neutral charge
cannot be affected by either the positive or negative charge. Yet
in nature, any neutral matter has its neutral charge. For example,
the congruency of neutrons to form a neutron star depends not
only on gravity; otherwise, non-neutron particles (e.g., protons,
electrons, anti-protons, anti-electrons, and positrons) would also
be encompassed by gravity. But such is not the case. Thus, it can
be concluded that the formation of a neutron star depends on the
neutral charge, which is responsible for neutron binding energy.

As discussed in Part IX, within the same metallic element,
one may find different numbers of neutrons, neutron-deficient
or neutron-rich isotopes, or neutron-electron interactions caused
by neutron-proton interactions. All of these cause the charges'
moment of momenta to create various crystalline structures. For
example, α-Fe, δ-Fe, and β-Zr form body-centered cubic (bcc)
crystals; γ-Fe and β-Co form face-centered cubic (fcc) crystals;
while α-Zr and α-Co form hexagonal closest-packed (hcp) crystals.

A neutron's neutral-charge features of high frequency, free oscillation, and quite low amplitude account for its strong binding energy. Thus, one may expect that the origin of the mu-particle is a pack of neutral-charge-field energy. The neutral charge field is a major binding force, as is the color (strong) force; while mu-particles are minor binding forces.

Section 2.

Time and Particle Universe

Time is truly a relative measurement. As is already known, traveling at higher speeds causes time to slow. In a high-voltage field, a clock pendulum swings more slowly; with strong gravitational force, time slows. In a strong magnetic field, time slows, but speeds up in the space between a strong magnetic field and a gravitational force field, which somehow cancel each other. All materials tend toward entropy and the weakening of energy. All combinations of constructions in this material universe are related to the three electrical charges.

Apart from the brane-structured gravity field, this universe consists of particles, including photons, neutrinos, and anti-neutrinos; thus it can be called a particle universe. If affected by the neutral charge field, the speed of the orbiting electrons will slow as they start to oscillate; if affected by inertia's acceleration, uniform speed, self-spin, and surrounding spiral spin or orbiting, the electron spin speed will slow. Therefore, anything that affects the charges and speed of particles will cause their time phenomenon to change. Thus, time in this particle universe is absolutely relative.

The cause of time's relativity is that every particle, even including the point particle, has its outer neutral field. If the outermost curvature degrees of two particles in close proximity are in the same direction as the curvature, they connect. Conversely, if the outermost curvature degrees of the two particles differ, they repel each other. Particles' self-spin, orbiting spin, and even oscillation frequency are slowed by the influence of any type of speed (e.g., uniform, acceleration, and spinning motion; repelling field). When this occurs, the particles' physical and chemical forces weaken, which causes time to slow. In such cases, minute particle radiations and packs of energy-emitting quantities and ratios will be reduced.

In the case of red or violet shifts, the respective intervals between two packs of quanta being emitted are either slower or faster than normal. It is not that the color has changed; rather, the interval frequency has either been delayed (red shift) or accelerated (violet shift). Therefore, both cases are related to time, which is directly related to the speeds of particles' motion. One might then consider that such a universe is constructed of particles in motion, along with time (i.e., measurement of variations in speed).

The structure of many parallel universes is not self-spin or orbiting particles. While such higher- and lower-layer universes are still composed of particles, their structure is virtual rather than material. Because there is no spin or orbiting speed, the measurement of speed relative to time is unnecessary. Thus, the concept of time cannot be established. Although these layer universes retain the cause-and-effect principle, cause and effect are not firmly linked, meaning there is always room for variance; thus, even predictions can change because any force, whether mind or matter, can affect the course or outcome.

Section 3.

Black Holes

It is the personal assumption of this author is that the non-particles existence in this particle universe is the gravity field. Every particle, no matter how gigantic or minute, is enwrapped in its neutral field, which can be combined into one neutral field with the same curvature. Therefore, the entire universe—encompassing all solar and galaxy systems—are turning or orbiting. But not all galaxies turn on the same plane. When two that turn on different planes meet, they produce massive radiation particles and energy; most of the matter produced will form into an even larger galaxy. However, two turning on the same plane in the same direction will repel each other strongly at the point where they touch. In this way, galaxy systems, layer by layer, have been pushed ever farther away. Thus, the outermost galaxies spin quite rapidly. Indeed, the universe scatter speed is shockingly fast.

In this case, should we not ask whether, subsequent to the Big Bang, the universe would eventually scatter away to never return. Here the author refers to the gravity field feature discussed in Part IV; that is, when the gravity field is spun ever faster, it flattens; it does not weaken in the center of the imaginary axle. Owing to gravity spin, the structure of the universe becomes flattened, with opposite top and bottom gravitational branes pushed thin to the extremity, at which point they are pulled back by two forces. First, the gravity force itself contracts to pull back the two-sided gravity branes from their expanded, flattened status. Second, a repellent force pushes two facing planes of the universe back into their original volume (i.e., dilation).

Thus, the turning back of the universe from the flattened status means that all galaxies, photons, and neutrinos are pulled back toward the center. But because the universal spin inertia is still present, the contracting spin is in the opposite direction.

During the contracting process, as the scattered materials of the galaxy move closer to the center, the universe increasingly changes from a flattened rounded shape into a three-dimensional, elliptical universe, thereby causing the density of matter to increase and accelerate. But the universe still cannot become a true sphere because the extremely dense matter and velocity will generate more power than nuclear fusion. This explosion of matter will cause a series of chain reactions of matter explosions spun to the center of the universe. Simultaneously, due to the control of the gravity field, once one side of the galaxy explodes along the turning arm, another spiral arm will usually explode along the opposite side of the galaxy. Thus, the great material explosion may not have only one exploding arm since this would tip over the great gravity field of the universe; moreover, the energy of only one arm is not that great.

When several material arms explode, a counter-explosion force in the center of the galaxy causes the center spin to accelerate more and friction to grind faster; that is the fastest whirl in the center. Here many "disembodied" atoms release all types of matter (e.g., neutrons, protons, neutrinos, electrons, and mesons). In the center acceleration, the force from everywhere in the center hole from two sides push out a positive charge similar to an axle, leaving only an empty hole still spinning, with the surrounding photons spinning along the edge: This is a black hole.

A black hole is empty. It does not contain matter condensed to the extremity, known as singularity. No. A black hole is indeed empty. However, it is surrounded by quite condensed matter; and other materials, attracted by the faster spin, are absorbed into the outer black hole.

At the time the Big Bang occurred, countless matter outside the black hole was journeying back toward the center hole. The super-high velocity of their return speed encountered the exploding material of the spinning arms. Because they moved in opposing directions, the impact speed accumulated. Thus, elements with high atomic numbers are not generated by stars or simply by a planet. Rather, they are created by counter-directional impacts.

It is known that some material found in space is older than the time of the Big Bang, suggesting that this event occurred before all material had returned to the center. Big Bang; black hole: They repeatedly begin and end the grand recycling; different galaxies have smaller big bangs and recycling, and the recycling of even smaller material explosions may occur to form small black holes outside the center of the galaxy.

Section 4.

Time Recycling

When will the time arrive for this particle universe to recycle all material back into its great black hole?

Many existing galaxies are not on the plane circle of this universe. They have different spin direction and angles. The flattening out of all galaxies onto the same plane requires waiting quite a long time since the universe is still expanding. Once extremely bright light can be observed coming from all directions in the sky—meaning that all photons will have been compressed into extremely condensed gravity branes—the time will have arrived for our universe to begin contracting back into a three-dimensional sphere. That is, waiting for all matter to journey back to the universal black hole for another Big Bang event will take as long as it did for the universe to extend out to the extremity.

Section 5.

Reflections

In all particles, the total positive charge is generally greater than the negative charge; the negative charge, which is the universal expanding force, will finally be called back by the positive charge. At the same time, the positive charge cannot prevent the negative charges from initiating another expansion. Thus, the universal total positive charge cannot shut down the universe forever. The extra positive charge may have the possibility of connecting either upward or downward to larger or smaller universes. Thus, countless universes can be linked.

If the structures of these larger and smaller universes are similar to that of our material universe, then they will never end, and their material nature, light prism, and many other features could be repeated in a variety of ways.

The higher or lower universal is outside the mathematical concept and thus a mathematical dimension universe (e.g., the fifth or sixth dimension) cannot be calculated. What can be calculated are connecting universes, set by set. Of course, they are without limit.

Some universes are constructed by spiritual force into point particles. In such universes, speed is without limit and thus time has no limit. Living there, one would be able to feel the past or the future simply as a presentation of one's mind. And one another's thought presentations would exist without interference.

Thus, the concepts of large and small or close and remote would have no meaning. But even this degree of freedom would extend only to the limits of that universe's layer. Moving from one layer universe to another, the number of dimensions may vary.

The thoughts expressed above are this author's own reflections. It is up to the reader to recognize different ways to further develop them for our better future.

Epilogue

PERFECT EQUILIBRIUM OF POSITIVE AND NEGATIVE charges in the universe would mean that matter would be locked without motion or that the encounter of two charges would change completely into quanta. But the three charges in this universe are not equally balanced; rather, each has its unique function and force. Therefore, motion is perpetual and is always circulating.

Furthermore, the finale of this universe is not entropy. On the contrary, since motion is always in action, the conditional existence of matter may have its cycle, meaning that time also cycles.

Moreover, with regard to gravity, the three charges have different formations, variations, and volume forces. Gravity simultaneously contains both the enwrapping and repellent forces, which cause the natural cycle of this material universe.

Finally, this book has focused particular attention on metals (Part IX) since using gravity's repellent force requires the development of an alloy that will not bend under any gravity field.

This book has presented the philosophical thoughts of the author, who initially considered a broad range of current data and multiple perspectives, along with known facts and observations, as a foundation for explaining the past and extrapolating the future.

Afterword

Dear Reader,

At this present moment in history, we find ourselves confined within the domain of time and space in this material universe. But we may expect that the existence of life is infinite. Even within our material universe, this author reckons that at least seven dimensions are functioning. Awareness of this concept reminds us that human beings still have much to develop.

Around our small globe, many individuals have their unique experiences, knowledge, and developments that will benefit the lives of all human beings and other lives in the future. Along the way, we mutually affect each other's development. This author has learned much from the wisdom of our great scientists, and takes this opportunity to extend his sincere respect to them.

It is true that this book substantially negates the theory of general relativity, which assumes that gravity depends on curved space. Yet this author maintains his deepest respect and gratitude to Albert Einstein, not only for his great scientific achievements, but especially for his extraordinary kindness and lack of egoism. Such unassuming qualities truly render one worthy of enduring respect and love.

Not only are we learning along the right course; we are also learning from our mistakes. Thus, this book too may need correcting. All the theories presented herein were developed from this author's pure thoughts; no means were available to do experimental research. Thus, I, the author, now leave it up to you, the reader, to discover the practical applications of the book's contents.

With my warm regards and deep concern for the future well-being of our planet and her people,
Huang, Hung C. Henry
Virginia, September 2010